A Workbook of
Electrochemistry

A Workbook of
Electrochemistry

John O'M. Bockris
Professor of Physical Chemistry
The Flinders University
Adelaide, Australia
and

Ronald A. Fredlein
Lecturer in Physical Chemistry
The University of Newcastle
New South Wales, Australia

PLENUM PRESS · NEW YORK — LONDON · 1973

Library of Congress Catalog Card Number 72-83606
ISBN 0-306-30590-9

© 1973 Plenum Press, New York
A Division of Plenum Publishing Corporation
227 West 17th Street, New York, N. Y. 10011

United Kingdom edition published by Plenum Press, London
A Division of Plenum Publishing Company, Ltd.
Davis House (4th Floor) 8 Scrubs Lane, Harlesden,
London NW10 6 SE, England

Printed in the United States of America

PREFACE

In this book, the objective has been to set down a number of questions, largely numerical problems, to help the student of electrochemical science. No collection of problems in electrochemistry has previously been published. The challenge which faces the authors of such a book is the breadth of the material in modern electrochemistry, and the diversity of backgrounds and needs of people who may find a "problems book" in electrochemistry to be of use.

The general intention for Chapters 2–11 has been to give the first ten questions at a level which can be dealt with by students who are undergoing instruction in the science of electrochemistry, but have not yet reached graduate standard in it. The last two questions in Chapters 2–11 have been chosen at a more advanced standard, corresponding to that expected of someone with knowledge at the level of a Ph.D. degree in electrochemistry.

Chapters 1 and 12 have a different character from the other chapters in the book. Chapter 1 is for the generalist who has some knowledge of the impending developments and rapid changes currently going on in electrochemistry, which results from the realization that we cannot continue with a fossil-fueled economy, and must turn increasingly to one which centers itself upon atomic energy and electricity. Chapter 12, on the other hand, presents a broad spectrum of problems in which no reservations have been made about the knowledge of the student. The questions in Chapter 12 should be answerable by an experienced graduate in electrochemical science—one who regards as his province the whole field, with its widespread applications.

In Chapters 2–11, there has been a further breakdown of the problems. The first two or three problems are particularly simple and generally need only some direct application of one or two equations.

The presence of simple, direct questions of this character seems to be necessary to help the student obtain familiarity with the units, symbols, and basic ideas of the field. Any student who finds continued difficulties in obtaining satisfactory answers to Problems 1 and 2 of Chapters 2–11 is not sufficiently familiar with the basic concepts of modern electrochemistry and should undertake further course work or individual study before proceeding.

A word must be said about the use of the book in courses of electrochemistry, electroanalytical chemistry, and electrochemical engineering. The book has been written predominantly for students in electrochemical science, i.e., that broad interdisciplinary field concerning the electrical properties of interfaces, and their contributions to the fields of metallurgy, energy conversion and storage, electrosynthesis, electrobiology, and so on. Students who wish to specialize in electroanalytical chemistry or electrochemical engineering have to go through a fundamental course in modern electrochemical science (which must of course be oriented in terms of quantum mechanics and solid state physics) as the foundation of their specialization. After such a fundamental course, they should be able to do at least Problems 1–10 of Chapters 1–11.

The time needed to work out most of the problems in Chapters 2–11 varies from a few minutes each for Problems 1 and 2 (so long as the basic material is well understood beforehand and the relevant formula immediately recognized), to times of the order of one to two hours each for the Problems 3–10, to times of the order of a day for Problems 11 and 12. These latter problems are intended as "take-home exam" type problems, and the availability of libraries and computing machines is assumed.

With respect to the necessary data, this has been given in Problems 1–10 of Chapters 1–11. In Problems 11 and 12 of Chapters 1–11, and in all problems of Chapter 12, little or no information has been given, to simulate the actual conditions of problem-solving in the laboratory or office.

Books of electrochemical constants are readily available and we may here mention the classic book by Latimer, *Oxidation Potentials* (2nd ed., 1952). Also, *Electrochemical Data* by Conway (1952) and the *Handbook of Electrochemical Constants* by Parsons (1959). Volumes 9a and 9b of *Physical Chemistry* by Eyring, Henderson, and Jost (1970) have many tables of constants, as do the books mentioned below.

For a time, a great paucity of books on modern electrochemistry existed, but the field has had the benefit, in the last few years, of several new books. Of these must be mentioned *Electrode Processes* by B. E. Conway (1965), *The Double Layer and Electrode Kinetics* by P. Delahay (1967), *Electrochemical Kinetics* by K. Vetter (1967), *Modern Electrochemistry* by J. O'M. Bockris and A. K. N. Reddy (1970), the two volumes of electrochemistry in the *Physical Chemistry* volumes mentioned above, and *Electrochemical Science* by J. O'M. Bockris and D. Drazic (1972).

Solutions have been provided to about half of the numerical problems of Chapters 2–11. No attempts have been made to give solution to the problems of Chapter 12. Here, the discussion of a full answer might sometimes give rise to an original publication and the "correct" answer to some of the applied problems may be regarded as a matter of options and be subject to changing social and economic pressures.

This book was written in the Electrochemistry Laboratory of the University of Pennsylvania. The authors acknowledge the assistance of many of their co-workers, especially: Dr. Bonciocat (Chapters 4 and 7), Dr. Buck (Chapter 6), Dr. McHardy (Chapters 9 and 10), Mr. O'Grady (Chapter 11), Dr. Paik (Chapters 2 and 3), and Mr. Sen (Chapters 5 and 8).

Adelaide
Newcastle
August 1972

J. O'M. Bockris
R. A. Fredlein

CONTENTS

CHAPTER 1

GENERAL QUESTIONS ABOUT ELECTROCHEMISTRY

1 Make up three definitions of electrochemistry. Discuss the appropriateness of the definitions to the future applications of electrochemistry. Do you think the field should embrace both the physical chemistry of ions in solution *and* the properties of charged interfaces, or be limited to the latter?

What about analytical chemists who devote themselves to the development of new electrochemical methods in analysis and the underlying electrochemistry? Should they be called electrochemists? What of technicians who work in plating shops? What of academics who specialize in quantum mechanical calculations of electrochemical kinetics? What of men who work at the electrochemistry of surgery, or people who design anticorrosion systems for ships? Should they *all* be called electrochemists or electrochemical scientists, or what should they be called?

2 Outline the subjects grouped under the heading of "materials science." What fields of endeavor and what applications does it embrace? Mention several aspects of materials science which rest upon electrochemical science.

3 Attempt to rationalize, in terms of the thermodynamics of ions in solids and in the liquid phase, the statement: "All surface problems involving one conducting phase in contact with another conducting phase have properties which can be discussed in terms of electrochemical theory."

| 4 | What evidence exists to support the statement: "Some heterogeneous chemical reactions occurring on solids in contact with gases may in fact have an electrochemical mechanism?" Illustrate the type of mechanism which may be involved.

| 5 | The theory of electron emission from metals to a vacuum has been treated classically. Why is it that for electrochemical reactions only a purely quantum mechanical treatment is significant? Make numerical calculations to illustrate your answer.

| 6 | Mention a dozen examples of practical phenomena in which electrochemical science is involved, and describe what that involvement is. Give the main reason why electrochemical science has such a wide involvement in many practical sciences.

| 7 | Consider the future of the field of electrochemistry. Should it continue as a branch of physical chemistry, and, at universities, should academics interested in contributing to the field work in chemistry departments? Or would you advocate the growth of a new discipline in universities ("electrochemical science") outside departments of chemistry? Discuss the pros and cons of establishing separate departments of electrochemical science and draw a conclusion.

| 8 | To what degree can electrodic reactions be thought of as an aspect of heterogeneous catalysis? To what extent may heterogeneous catalysis be looked at in electrodic terms?

| 9 | Discuss three major antipollutional developments, affecting the whole community, which you would expect to see accomplished by, say, 2000 A.D., and which depend upon the development of electrochemical technology.

| 10 | The German physical chemist Nernst must be reckoned as one of the greatest chemists of all time. Estimate his effect on

electrochemistry and the results of that influence on the quality of life in the 21st century.

11 Weigh the statement: "Electrochemical science is a new applied science." How do you think electrochemical science compares with materials science, energy conversion, and communications, as a new science? Estimate the circumstances which led to the identification and "coming out" of these new sciences. Consider in detail the three new sciences, and estimate the date of their identification and decide which early discoveries and formulations constitute their roots. Estimate the time which it took for these sciences to become accepted in their own right as university topics.

12 Increasing pollution of the atmosphere and the waters, diminishing reserves of raw materials and fossil-fuels, and increasing world population are some of the problems which beset man's existence on this planet. Abundant electricity, at a cost of about one-fifth the present cost (in present dollars) is technologically feasible from very large breeder reactors. To what degree (and in what ways) would the picture of plenty for all mankind, held out by the advocates of breeder reactors, be dependent on the extensive development of a new electrochemical technology? To what extent are there alternative technological paths which do not involve electrochemistry, but which would be entirely free of pollutant emissions, in particular, carbon dioxide?

CHAPTER 2

ION–SOLVENT INTERACTIONS

| 1 | The experimentally determined heat of sublimation of rubidium iodide is $\overline{+}148.6$ kcal mole^{-1} and the corresponding heat of solution is $+6.2$ kcal mole^{-1}.

Calculate (a) the heat of hydration of the salt from the experimental data; (b) the same quantity, in terms of the Born model, utilizing the following data: $r_{Rb^+} = 1.48$ Å; $r_{I^-} = 2.16$ Å; $\epsilon_w = 78.3$ (25°C); $d\epsilon_w/dT = -0.356$ deg^{-1}.

| 2 | Using the data of Problem 1 and $r_{Cl^-} = 1.81$ Å, calculate the entropy of hydration of the Cl$^-$ ion, in terms of the Born model.

| 3 | The determination of the absolute heat of hydration of the proton by Halliwell and Nyburg (1963) ($\Delta H_{H^+-h} = -266$ kcal mole^{-1}) permits the absolute heat of hydration of other ions to be

TABLE 2.3.1.   Integral Heats of Solution of NaCl at Various Concentrations*

10^3 Concentration (molal)	ΔH_s, cal mole^{-1}
1.45	952 ± 33
2.88	931 ± 18
5.31	941 ± 15
6.39	945 ± 47
8.98	959 ± 2
9.93	937 ± 6
20.56	965 ± 5

* Data from Criss and Cobble (1961).

5

calculated. From the integral heats of solution of NaCl at 25°C
(Table 2.3.1), calculate the absolute heats of hydration of Na$^+$ and Cl$^-$.
The relative heat of hydration of Cl$^-$, $\Delta H^{rel}_{Cl^- - h}$, is -347.5 kcal mole^{-1}
and the heat of sublimation of NaCl is $+184.7$ kcal mole^{-1}.

---

4   Consider a dilute solution of NaCl at 25°C. Calculate for each
ion: (a) the Born heat of solvation; (b) the ion–dipole heat;
(c) the ion–quadrupole heat; (d) the induced ion–dipole interaction;
(e) the heat of structure breaking in the solvent.

$$r_w = 1.38 \text{ Å}; \qquad r_{Cl^-} = 1.81 \text{ Å}; \qquad r_{Na^+} = 0.95 \text{ Å}$$

The dipole moment of water, $\mu_w = 1.87 \times 10^{-18}$ esu cm.

The quadrupole moment $p_w = 3.9 \times 10^{-26}$ esu cm^2.

The polarizability $\alpha_w = 1.46 \times 10^{-24}$ cm^3.

Make clear what assumptions are made for each calculation, and
state what you think of the accuracy. Make an estimate of the heat of
solvation of Na$^+$ and Cl$^-$ ions arising thus, and then discuss the likely
meaning of discrepancies with the experimental values determined in
Problem 3, in terms of modelistic considerations.

---

5   In Table 2.5.1, data are given on the adiabatic compressibility
of solutions of certain salts. The adiabatic compressibility of
water at 25°C is 44.730×10^{-6} bar^{-1}.

**TABLE 2.5.1. Adiabatic Compressibility and Density (ρ)
of Electrolyte Solutions as a Function of Concentration (C) at 25°C**

Salt	C, moles liter^{-1}	ρ, g ml^{-1}	Adiabatic compressibility, $\times 10^6$ bar^{-1}
NaI	5.046	1.5617	28.783
	2.053	1.2290	36.831
	1.013	1.1120	40.838
	0.100	1.0086	44.278
	0.0500	1.0028	44.481
CaCl$_2$	4.053	1.3244	22.900
	2.019	1.1673	30.940
	1.005	1.0843	36.907
	0.101	1.0062	43.823
	0.0510	1.00176	44.224

Utilizing the relation of Passynski (1938) between the number of incompressible water molecules around an ion in solution and the adiabatic compressibility of the solution, calculate the total hydration number of NaI and $CaCl_2$ at each concentration. Discuss to what extent the result you obtain represents the hydration number of the salt concerned. Define carefully the hydration number to which you refer.

6 Ionic vibration potential measurements permit the calculation of

$$(t_+ n_+/| z_+ |) - (t_- n_-/| z_- |) \qquad (2.6.1)$$

for an electrolyte in solution, where n_+ and n_- are the individual solvation numbers of the cation and anion, respectively, and t_+ and t_- are the transport numbers at infinite dilution. Using the equations presented by Zana and Yeager (1967) and the data of Tables 2.6.1 and 2.6.2 together with the results of Problem 5, calculate the individual ionic hydration numbers of the ions. The radii of hydrated ions, estimated by Nightingale (1959), are given in Table 2.6.3.

TABLE 2.6.1. Transport Numbers at Infinite Dilution of Ions in NaI and $CaCl_2$ Solutions

Electrolyte	t_+	t_-
NaI	0.395	0.605
$CaCl_2$	0.438	0.562

7 Consider the water in the first layer around an ion, e.g., Na^+, in aqueous solution. Write an expression for the total electrostatic interaction energy of the ion with a water molecule. Explain the physical meaning of all the terms. Deduce the electrostatic field strength at the center of the water molecule.

Discuss the basis on which one could conclude whether the water molecules are dielectrically saturated. Use the equations of Booth (1951), which relate dielectric constant to the field strength to estimate the degree of saturation of water in the first layer near the sodium ion. $r_{Na^+} = 0.95$ Å; $r_w = 1.38$ Å.

TABLE 2.6.2.   Ionic Vibration Potential Terms (ϕ_0/a_0)*
for NaI and CaCl$_2$ Solutions at Various Concentrations

Electrolyte	Concentration, M	$10^6(\phi_0/a_0)$, V cm^{-1} sec
NaI	0.05	−6.4
	0.10	−6.4
	1.0	−5.7
	5.0	−5.7
CaCl$_2$	0.05	1.1
	0.10	1.1
	1.0	1.0
	5.0	0.8

* ϕ_0 is the ionic vibration potential and a_0, the velocity amplitude of the solvent, arises from the equation $v_0 = a_0 \exp[-i(\omega t - \sigma x)]$ which describes the periodic velocity of the solvent. ϕ_0/a_0 is obtained directly from experiment.

TABLE 2.6.3.
Hydrated Radii of Some Ions

Ion	r_{i-h}, Å
Na$^+$	3.58
I$^-$	3.31
Ca^{2+}	4.12
Cl$^-$	3.32

8   Distinguish, in terms of a physical argument, the external and internal fields in a dielectric. Calculate the ratio of these fields for benzene. $\epsilon_{C_6H_6} = 2.274$.

In Kirkwood's theory of the dielectric constant of associated liquids, the factor g is usually taken as four. Draw a diagram to show how this factor arises from concepts of the structure of water. Stress the physical, rather than the formal, meaning.

What degree of applicability would you expect Kirkwood's theory to have to liquid ammonia?

9   Describe qualitatively why dielectric constants of ionic solutions become less with increasing concentration of dissolved ions.

The empirical relation

$$\epsilon_s = \epsilon_w - \delta C \tag{2.9.1}$$

where ϵ_s and ϵ_w are the dielectric constant of an electrolyte solution and water, respectively, C is the concentration, and δ a constant, has been found to be obeyed up to concentrations of ~ 1 M. By considering the solution as a suspension of spheres of dielectric constant ϵ_{ions} in a medium of dielectric constant ϵ_w, the relation

$$\frac{(\epsilon_s/\epsilon_w) - 1}{(\epsilon_s/\epsilon_w) + 2} = \rho \frac{(\epsilon_{ions}/\epsilon_w) - 1}{(\epsilon_{ions}/\epsilon_w) + 2} \tag{2.9.2}$$

has been deduced (Fricke, 1924). Here, ρ is the volume fraction of the solute. Derive (2.9.1) from (2.9.2) and show that

$$\delta = 1.5 \left[V_{ions} \frac{\epsilon_w - \epsilon_{ions}}{1000} + V_w \frac{\epsilon_w - \epsilon_{w\infty}}{1000} n_h \right] \tag{2.9.3}$$

where V_{ions} and V_w are the actual molar volumes of solute and water, respectively, $\epsilon_{w\infty}$ is the deformation polarization of water, and n_h is the total primary hydration number. Make clear the assumptions and approximations of your derivation. Could the hydration number of an electrolyte be obtained from (2.9.1) and (2.9.3)?

10 | What is the definition of the "primary solvation number"? How does it differ from the "solvation number"? Distinguish clearly between the coordination number and the primary solvation number of an ion in solution.

It has been suggested that the essential difference between solvation number and coordination number resides in the time an ion remains stationary in a given site in solution, τ_1, compared with the time it takes to orient a water molecule out of the surrounding water structure and into a position of maximum interaction with the ion, τ_2. If $\tau_2 < \tau_1$, the water molecule concerned becomes part of the hydration shell, and counts in the solvation number; if $\tau_1 < \tau_2$, the water molecule concerned remains only a coordination water.

From the equation

$$\Delta^2 = 2Dt \tag{2.10.1}$$

where Δ is the average distance traversed in one hop, calculate τ_1. (Take $D \simeq 2 \times 10^{-5}$ cm² sec⁻¹.) What sort of a value would you suggest for τ_2 ? How would you go about making (a) a quick estimate of τ_2, (b) a more thorough estimate of τ_2 ?

Utilize approximate values of τ_2 to consider the rationality of the distinction between solvation number and coordination number.

11 Some experimental entropies of hydration of ions are listed in Table 2.11.1.

TABLE 2.11.1. Entropies of Hydration
of Individual Ions

Ion	$S°$, cal deg⁻¹ mole⁻¹
H^+	-31.0
Na^+	-26.3
Rb^+	-15.6
Tl^{3+}	-16.4
Sr^{2+}	-56.6
Cd^{2+}	-66.5
Br^-	-14.4

Give an interpretation of the negative sign of this entropy in terms of continuum theory.

Suppose that the model for the region near an ion is given as follows: The coordination number varies from ion to ion and for Na^+ it may be taken as six. The solvation number (the number of water molecules oriented toward the ion in positions of maximum interaction) is four for Na^+. The second layer consists of a structure-broken region and outside the second layer the structure is that of the bulk.

Calculate the entropy of Na^+ in the gas phase. Formulate the various structural, structure-breaking, and Born contributions to the solvational entropy, allowing for the difference of coordination number and solvation number.

See how near your calculations can bring you to the measured values of Na^+ and other ions listed in Table 2.11.1 (you can be satisfied with discrepancies up to about 33 %).

12 Work out the quadrupole interaction between an ion and a water molecule and show that, for a positive ion, it is given by

$$U = -\frac{z_i e_0 \mu_w}{r^2} + \frac{z_i e_0 p_w}{2r^3} \qquad (2.12.1)$$

where p_w is the quadrupole moment of water, μ_w is the dipole moment, and r is the distance from the center of the ion to the center of the water molecule when the ion and water molecule are in contact, and the water molecule is oriented for maximal interaction. In your deduction, be very careful to make a clear and diagrammatic definition of the quadrupole moment. Show that it consists of several contributions of the type of x^2, where x is a distance.

Calculate the error introduced into the calculated heat of hydration of a univalent cation and anion by neglecting the quadrupole character of water. Make an estimate of the uncertainty of the heats of hydration of individual ions calculated from the ion-quadrupole model.

ANSWERS

1 (a) The heat of solution ΔH_s of an ionic crystal is equal to the heat of dissociating the crystal into its individual ions (i.e., the heat of sublimation ΔH_{subl}) plus the heat of hydration of the ions ΔH_h. Thus,

$$\Delta H_s = \Delta H_{\text{subl}} + \Delta H_h \qquad (2.1.1)$$

Upon substituting the numerical values, the heat of hydration of RbI is found to be

$$\Delta H_{\text{RbI}-h} = 6.2 - 148.6 \text{ kcal mole}^{-1}$$
$$= -142.4 \text{ kcal mole}^{-1}$$

(b) From the Born equation, the heat of hydration of an ion i, ΔH_{i-h}, of radius r_i and charge z is given by

$$\Delta H_{i-h} = -\frac{N_A z^2 e_0^2}{2r_i}\left(1 - \frac{1}{\epsilon_w} - \frac{T}{\epsilon_w^2}\frac{\partial \epsilon_w}{\partial T}\right) \qquad (2.1.2)$$

where N_A is Avogadro's number and ϵ_w is the dielectric constant of water. Substituting given values yields

$$\Delta H_{i-h} = - \frac{6.03 \times 10^{23} \times (4.8 \times 10^{-10})^2}{2r_i} \left(1 - \frac{1}{78.3} + \frac{298 \times 0.356}{(78.3)^2}\right)$$

$$\Delta H_{i-h} = - \frac{6.54 \times 10^4}{r_i} \; \text{esu}^2 \; \text{cm}^{-1} \; \text{mole}^{-1}$$

$$= - \frac{6.54 \times 10^4}{r_i} \; \text{ergs mole}^{-1}$$

when r_i is expressed in centimeters,

$$\Delta H_{RbI-h} = \Delta H_{Rb^+-h} + \Delta H_{I^--h}$$

$$= -6.54 \times 10^4 \left(\frac{1}{r_{Rb^+}} + \frac{1}{r_{I^-}}\right)$$

Substituting the given values for r_{Rb^+} for r_{I^-} yields

$$\Delta H_{RbI-h} = -7.45 \times 10^{12} \; \text{ergs mole}^{-1}$$

$$= -178 \; \text{kcal mole}^{-1}$$

| 3 | The relative heat of hydration of an ion i, ΔH_{i-h}^{rel}, is related to the absolute heat of hydration by |

$$\Delta H_{i-h} = \Delta H_{i-h}^{rel} - \Delta H_{H^+-h}$$

From the given data

$$\Delta H_{Cl^--h} = -347.5 + 266 \; \text{kcal mole}^{-1}$$

$$= -81.5 \; \text{kcal mole}^{-1}$$

The heat of solution of sodium chloride, ΔH_{NaCl-s}, can be found by extrapolating the integral heats of solution to infinite dilution; a reasonable straight line plot, suitable for extrapolation, being obtained by plotting against the square root of the molality. From Fig. 2.3.1,

$$\Delta H_{NaCl-s} = 0.91 \; \text{kcal mole}^{-1}$$

$$\Delta H_{NaCl-h} = \Delta H_{NaCl-s} - \Delta H_{NaCl-subl}$$

$$= 0.91 - 184.7 \; \text{kcal mole}^{-1}$$

$$= -183.8 \; \text{kcal mole}^{-1}$$

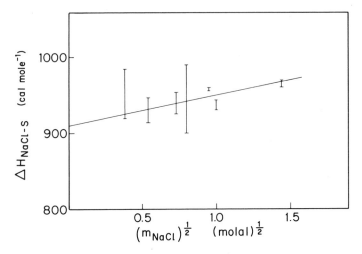

Fig. 2.3.1. Determination of the integral heat of solution of sodium chloride at infinite dilution.

Since

$$\Delta H_{\text{NaCl}-h} = \Delta H_{\text{Na}^+-h} + \Delta H_{\text{Cl}^--h}$$

$$\Delta H_{\text{Na}^+-h} = -183.8 + 81.5 \text{ kcal mole}^{-1}$$

$$= -102.3 \text{ kcal mole}^{-1}$$

⑤ The Passynski (1938) relation between the total hydration number of an electrolyte n_h and the compressibilities β_{soln} and β_{solv} of solution and solvent, respectively, is

$$n_h = (N_1/N_2)(1 - \beta_{\text{soln}}/\beta_{\text{solv}}) \qquad (2.5.1)$$

where N_1 is the number of moles of solvent molecules in the solution containing N_2 moles of salt.

For one liter of solution,

$$N_2 = \text{molar concentration}$$
$$N_1 = (1000\rho - N_2 M_2)/M_1$$

where M_2 is the molecular weight of the solute and $M_1 = 18.02$ for water.

TABLE 2.5.2.   Calculation of the Total Primary Hydration Number n_h

	C, M	$1 - \beta_{soln}/\beta_{solv}$	N_1	n_h
NaI	5.046	0.3565	44.7	3.16
	2.053	0.1766	51.1	4.40
	1.013	0.0870	53.3	4.61
	0.100	0.0100	55.1	5.58
	0.0500	0.00557	55.2	6.16
CaCl$_2$	4.053	0.4880	48.5	5.84
	2.019	0.3083	52.3	7.99
	1.005	0.1749	54.0	9.40
	0.101	0.0203	55.2	11.1
	0.0510	0.0113	55.3	12.3

Table 2.5.2 summarizes the results of the intermediate steps of the calculation with the final result, n_h, in the last column.

The calculation is based on the assumption that the solvating water molecules are totally incompressible while all other water molecules retain their normal compressibility. These assumptions have been reviewed by Bockris and Saluja (1972) and they conclude that the solvating water molecules are incompressible but that coordinating nonsolving water molecules are partly ($<30\%$) compressed. This leads to the conclusion that n_h calculated from (2.5.1) is too large by about 0.5 water molecules.

---

7 The interaction energy consists of the ion–dipole interaction term and the ion–quadrupole interaction term. For a positive ion, it becomes

$$E = - \frac{e_0 \mu_w}{(r_i + r_w)^2} + \frac{e_0 p_w}{2(r_i + r_w)^3} \qquad (2.7.1)$$

where μ_w and p_w are the dipole moment and quadrupole moment of the water molecule, respectively.

The first term is the interaction energy of a dipole of water with the field due to the ion. The second term, the quadrupole–ion interaction energy, stems from the fact that the distribution of charge in the water molecule is such that the electrostatic interaction energy between the charges in a water molecule and the ionic charge cannot be accounted for by only the dipole–field energy. The quadrupole energy accounts for most of this difference. The field due to the ionic charge e_0 is

$$X_{ion} = e_0/(r_i + r_w)^2$$
$$= (4.8 \times 10^{-10})/(2.33 \times 10^{-8})^2$$
$$= 0.86 \times 10^6 \text{ statvolt/cm} \qquad (2.7.2)$$

The expression for the dielectric constant is (Booth, 1951)

$$\epsilon = \epsilon_0 + \frac{\alpha \pi N_0 (\epsilon_0 + 2) \, \mu_w}{X_{ion}} \, \mathscr{L} \left[\frac{\beta \mu_w (\epsilon_0 + 2) X_{ion}}{kT} \right] \qquad (2.7.3)$$

where $\alpha = 28/3\sqrt{73}$ and $\beta = \sqrt{73/6}$. X_{ion} is the field due to the ion in the absence of a medium and N_0 is the number of molecules per cubic centimeter. It should be noted that ϵ_0 is the result of both atomic and electronic polarization and is equal to six [Rampolla *et al.* (1959)].

$\mathscr{L}(x)$, the Langevin function of x, is given by

$$\mathscr{L}(x) = \coth x - 1/x$$

With $X_{ion} \simeq 10^6$ statvolt cm^{-1}, the argument of $\mathscr{L}$ becomes

$$x \simeq \sqrt{73/6} \, (1.8 \times 10^{-18})(8 \times 10^6)/(4.2 \times 10^{-14})$$
$$\simeq 13.5 \times 10^2$$

Thus

$$\mathscr{L}(x) \simeq 1$$

and

$$\epsilon = \epsilon_0 + [\alpha \pi N_0 (\epsilon_0 + 2) \, \mu_w/X_{ion}]$$

Substituting numerical values yields the dielectric constant in the first layer:

$$\epsilon = \epsilon_0 + \frac{28\pi \times 6.023 \times 10^{23} \times 8 \times 1.87 \times 10^{-18}}{3 \times (73)^{1/2} \times 18 \times 0.86 \times 10^6}$$
$$= \epsilon_0 + 1.5$$

Since the contribution from orientation polarization to the dielectric is small (1.5), this means that the dielectric saturation of water in the first layer around an ion is essentially complete.

9   Water molecules near an ion are under the influence of a strong electric field due to the ion; therefore, they are completely aligned to that field, fixed and insensitive to the external field. Hence, only the nuclear and electronic contributions to the dielectric constant are important for the hydrating water molecules. The dielectric constant arising from these contributions is ~6.

The water molecules in the secondary region are also affected by structure breaking, which leads to a smaller value of the number of water molecules which turn as a group and hence to smaller ϵ.

Rearranging (2.9.2) yields

$$\frac{[(\epsilon_s/\epsilon_w) - 1][(\epsilon_{ions}/\epsilon_w) + 2]}{[(\epsilon_s/\epsilon_w) + 2][(\epsilon_{ions}/\epsilon_w) - 1]} = \rho$$

and

$$\frac{[(\epsilon_s/\epsilon_w) - 1][(\epsilon_{ions}/\epsilon_w) + 2]}{[(\epsilon_s/\epsilon_w) + 2][(\epsilon_{ions}/\epsilon_w) - 1] - [(\epsilon_s/\epsilon_w) - 1][(\epsilon_{ions}/\epsilon_w) + 2]} = \frac{\rho}{1 - \rho}$$

Rearranging gives

$$\frac{[(\epsilon_s/\epsilon_w) - 1][(\epsilon_{ions}/\epsilon_w) + 2]}{(\epsilon_{ions}/\epsilon_w) - (\epsilon_s/\epsilon_w)} = \frac{3\rho}{1 - \rho}$$

and since $\epsilon_{ions} \simeq 2$ while $\epsilon_w = 78.3$, we have

$$\epsilon_s - \epsilon_w = 1.5[\rho/(1 - \rho)](\epsilon_{ions} - \epsilon_s) \qquad (2.9.4)$$

Rearranging (2.9.4) gives

$$\epsilon_s = \epsilon_w + [1.5\rho(\epsilon_{ions} - \epsilon_w)/(1 + 0.5\rho)] \qquad (2.9.5)$$

For solutions that are not too concentrated, $0.5\rho \ll 1$, where ρ is the volume fraction of the hydrated ion, and (2.9.5) becomes

$$\epsilon_s = \epsilon_w - 1.5\rho(\epsilon_w - \epsilon_{ions}) \qquad (2.9.6)$$

Equation (2.9.6) holds for ions that are not hydrated. For hydrated ions (2.9.6) becomes

$$\epsilon_s = \epsilon_w - 1.5\rho_h(\epsilon_w - \epsilon_{ions-h}) \qquad (2.9.7)$$

where ϵ_{ions-h} is the dielectric constant of hydrated ions and ρ_h, the volume fraction of hydrated ions, is given by

$$\rho_h = CV_{ions}/1000 + Cn_hV_w/1000 \qquad (2.9.8)$$

To a first approximation,

$$\epsilon_{\text{ions}-h} = f_i \epsilon_{\text{ions}} + f_w \epsilon_{w\infty} \qquad (2.9.9)$$

where f_i is the volume fraction of the ion in a hydrated ion and f_w is the volume fraction of restricted water in a hydrated ion. $\epsilon_{w\infty}$ is the dielectric constant of restricted water. Consequently, (2.9.9) can be written as

$$\epsilon_{\text{ions}-h} = \frac{V_{\text{ions}}}{V_{\text{ions}} + n_h V_w} \epsilon_{\text{ions}} + \frac{n_h V_w}{V_{\text{ions}} + n_h V_w} \epsilon_{w\infty} \qquad (2.9.10)$$

Substituting (2.9.8) and (2.9.10) into (2.9.7) yields

$$\epsilon_s = \epsilon_w - \delta C \qquad (2.9.1)$$

where

$$\delta = 1.5 \left[V_{\text{ions}} \frac{\epsilon_w - \epsilon_{\text{ions}}}{1000} + V_w \frac{\epsilon_w - \epsilon_{w\infty}}{1000} n_h \right] \qquad (2.9.3)$$

Because (2.9.9) is only an approximation and because the total primary hydration number n_h of the electrolyte is assumed to be independent of concentration, application of equations (2.9.1) and (2.9.3) would not constitute a reliable method of finding n_h.

CHAPTER 3

ION–ION INTERACTIONS

1 Calculate the radius of the ionic atmosphere in aqueous potassium bromide solution. $C_{KBr} = 10^{-3}$ mole liter^{-1}, $\epsilon_w = 78.3$.

2 Calculate the number of ions comprising the ionic atmosphere around a K^+ and also a Br^- in a solution of 10^{-3} M potassium bromide. What is the average distance apart of such ions?

3 Calculate the self-energy (kcal mole^{-1}) of Na^+ in dilute aqueous solution. Correspondingly, calculate the energy of the ionic atmosphere for the same ion, in 10^{-3} M sodium chloride solution. $r_{Na^+} = 0.95$ Å; $\epsilon_w = 78.3$.

4 Derive in two different ways (i.e., using the Debye treatment and the Guntelberg treatment) an expression for the reversible charging of an ion in a solution of constant ionic strength, taking into account the radius of the ion. Explain clearly the difference in the physical assumptions of the two methods. In deriving expressions for the free energy of ion–ion interactions in solution, what value is there in comparing the results obtained by using the two treatments of reversible ion charging?

5 Derive equations to relate the root mean square activity coefficient on the mole fraction scale to those on the molarity and molality scales. The root mean square activity coefficient of the ions in 3.2 M aqueous calcium chloride solution is 1.55. The density of the solution is 1.246 g ml^{-1}. Calculate the root mean square activity coefficient on the molality scale and on the mole fraction scale.

6 Explain fully and clearly what is meant by a standard state in problems connected with solutions. What standard states are frequently used for ionic solutions? Which is normally used? Is the standard state for the solute a hypothetical or a real state? Distinguish a reference state from a standard state in solution theory.

7 Utilize the Fuoss approach to ion pair formation to find the ionic association constant of NaCl (a) in water at 5 M; and (b) in benzene at 10^{-6} M. $\epsilon_w = 78.3$; $\epsilon_{C_6H_6} = 2.27$.

 8 In the calculation of the free energy of interactions of ions in solution, the mathematical approximation of linearizing the Boltzmann term is made, i.e.,

$$\exp(-z_i e_0 \psi_r / kT) = 1 - (z_i e_0 \psi_r / kT) \qquad (3.8.1)$$

where $z_i e_0 \psi_r$ is the Coulombic potential energy of an ion of charge $z_i e_0$ in an electrostatic potential ψ_r. Show that it is pointless to attempt a more rigorous solution by avoiding this approximation.

9 The spectroscopic determination of the association constant of copper sulfate in water is carried out by measuring the difference in optical densities of two solutions. Solution (1) contains $Cu(ClO_4)_2$, Na_2SO_4, and $NaClO_4$ to maintain constant ionic strength, while solution (2) contains $NaClO_4$ and the same amount of $Cu(ClO_4)_2$ as solution (1). Assuming the Beer–Lambert law, show that at wavelengths at which only Cu^{++} and associated $CuSO_4$ absorb,

$$D_1 - D_2 = C_{CuSO_4} l (\epsilon_{CuSO_4} - \epsilon_{Cu^{++}}) \qquad (3.9.1)$$

where D_1 and D_2 are the optical densities of solutions (1) and (2), C_{CuSO_4} is the concentration of associated $CuSO_4$, l is the length of the cell, and ϵ_{CuSO_4} and $\epsilon_{Cu^{++}}$ are the molar absorptivities of $CuSO_4$ and Cu^{++}, respectively.

The data in Table 3.9.1 pertain to three approximately constant ionic strengths I_1, I_2, and I_3. Calculate

$$K' = C_{CuSO_4}/C_{Cu^{++}}C_{SO_4^{--}} \tag{3.9.2}$$

at each ionic strength and roughly estimate the association constant $K = a_{CuSO_4}/a_{Cu^{++}}a_{SO_4^{--}}$, using Debye–Hückel theory with parameters A and B equal to 0.5115 and 0.3291×10^8 respectively.

TABLE 3.9.1. Optical Density $(D_1 - D_2)$*
for Various Solution Compositions†

$10^3 C_{Cu(ClO_4)_2}$, M	$10^3 C_{Na_2SO_4}$, M	$10^3 C_{NaClO_4}$, M	$D_1 - D_2$
I_1			
4.05	8.00	56.5	0.143
4.05	12.0	45.2	0.206
4.05	16.0	33.9	0.260
4.05	20.0	22.6	0.312
4.05	24.0	11.3	0.352
4.05	28.0	0.0	0.389
I_2			
5.00	4.74	42.6	0.134
5.00	7.10	36.4	0.187
5.00	9.46	29.3	0.247
5.00	11.83	22.7	0.293
5.00	14.21	17.3	0.340
I_3			
5.00	1.66	41.1	0.046
5.00	3.34	26.9	0.087
5.00	6.66	28.5	0.160
5.00	8.34	24.2	0.193
5.00	10.00	20.0	0.226

* $D_1 - D_2$ is the optical density of the solution of interest minus that of a blank.
† Data from Hemmes and Petrucci (1968).

10. Write elucidatory notes on the following:

(a) The reason for the term potential electrolyte.

(b) The theory which gives the range in which the Poisson–Boltzmann equation can be linearized.

(c) The significance of negative values of the ion size parameters.

(d) A relation which may exist between the water molecules counted in the Stokes–Robinson model for concentrated electrolyte solutions and the primary solvation number.

(e) Approaches by which individual activity coefficients may be estimated.

(f) The modelistic meaning of activity coefficients greater than one.

| 11 | Essentially, the theory of ionic solutions is successful up to about 0.005 N, whereafter there is a dependence upon empiricisms to obtain agreement with experiment. Review the rationale of these empiricisms. In particular, examine the validity of the Stokes–Robinson model for solutions of higher concentrations. How valid is the use of the ion–ion cloud interaction term in equations for the activity coefficients when applied to very concentrated solutions? Discuss the implications for models of the agreement with experiment of the log $\gamma_\pm$ against concentration relation which Stokes and Robinson obtain.

Use the data of Table 3.11.1 and the Stokes–Robinson relation to calculate the energy of the ion–ion cloud interaction. Examine the relative importance of this term in the Stokes–Robinson treatment of concentrated solutions of electrolytes and examine its concentration

TABLE 3.11.1. Data for Calculation of the Ion–Ion Cloud Interaction Energy for Concentrated Sodium Chloride Solutions at 25°C

Concentration (molal)	$\gamma_\pm$ (molal scale)	Osmotic coefficient (molal scale)	Total hydration number
0.1	0.778	0.932	6.6
0.5	0.681	0.921	6.1
1.0	0.657	0.936	5.6
2.0	0.668	0.983	4.7
5.0	0.874	1.192	3.8

dependence. Assume that the primary hydration number is the appropriate quantity to use in this calculation.

12 Synthesize the position of the 1970's on the physical aspects of the theoretical treatment of concentrated ionic solutions, paying attention to the work of Frank and Thompson (1960) and Frank (1966) on the dependence of log γ on $C^{1/3}$, and to that of Friedman (1971) on the cluster theory in electrolytes. Discuss possible links between these theories of concentrated aqueous solutions and models of simple molten salts associated with the names of Angel, Bloom, Rice, Richards, Riess, and Stillinger.

ANSWERS

1 Since the radial distribution of ionic charge in the ionic atmosphere attains its maximum value at a distance κ^{-1} from the central ion, where κ is the Debye reciprocal length, the radius of the ionic atmosphere is defined as κ^{-1}. Hence,

$$r = \kappa^{-1} = \left(\epsilon kT \big/ 4\pi \sum_i n_i z_i^2 e_0^2\right)^{1/2} \tag{3.1.1}$$

where n_i is concentration of the ionic species and ϵ is the dielectric constant of the solvent. The other symbols have their usual meanings. From a consideration of the units, κ^{-1} has the units centimeters when n_i is in ions per cubic centimeter, i.e. (with $C_{KBr} = 0.001M$),

$$n_{K^+} = n_{Br^-} = N_A C_{KBr}/1000$$
$$= 6.023 \times 10^{17} \text{ ions cm}^{-3}$$

Substituting this and given values into (3.1.1) yields

$$r = \left[\frac{78.3 \times 1.38 \times 10^{-16} \times 298}{2 \times 4\pi \times 6.023 \times 10^{17} \times (4.8 \times 10^{-10})^2}\right]^{1/2}$$
$$= 9.7 \times 10^{-7} \text{ cm}$$

⎡3⎤ The self-energy W of an ion in a medium of dielectric constant ϵ is approximated by the energy to charge a sphere of radius r_i equal to the crystallographic radius of the ion. Thus, for Avogadro's number of ions,

$$W = [(z_i e_0)^2 / 2 r_i \epsilon] \, N_A \tag{3.3.1}$$

Substituting numerical values yields

$$W = \frac{(4.8 \times 10^{-10})^2 \times 6.023 \times 10^{23}}{2 \times 0.95 \times 10^{-8} \times 78.3}$$

$$= 0.933 \times 10^{11} \text{ ergs mole}^{-1}$$

$$= 2.23 \text{ kcal mole}^{-1}$$

The energy of the ionic atmosphere is the extra chemical potential due to the ion–ion cloud interaction, $\Delta\mu_{i-I}$. From the Debye–Hückel theory,

$$\Delta\mu_{i-I} = -\frac{N_A}{2} \frac{(z_i e_0)^2}{\epsilon \kappa^{-1}} \tag{3.3.2}$$

where κ^{-1}, the Debye–Hückel reciprocal length, has been evaluated for a $10^{-3} \, M$ solution of a 1:1 electrolyte in Problem 1 as

$$\kappa^{-1} = 9.7 \times 10^{-7} \text{ cm}$$

Substituting numerical values in (3.3.2) yields

$$\Delta\mu_{i-I} = -\frac{6.023 \times 10^{23} \times (4.8 \times 10^{-10})^2}{2 \times 78.3 \times 9.7 \times 10^{-7}}$$

$$= -0.914 \times 10^9 \text{ ergs mole}^{-1}$$

$$= -2.18 \times 10^{-2} \text{ kcal mole}^{-1}$$

⎡5⎤ For an electrolyte that gives rise to ν_+ cations and ν_- anions in solution, the mole fraction x_i of the ionic species i is related to the molarity C and the molality m by

$$x_+ = \frac{0.001 C M_1 \nu_+}{\rho - 0.001 C M_2 + 0.001 C M_1 \nu} \tag{3.5.1}$$

$$= \frac{0.001 m M_1 \nu_+}{1 + 0.001 m M_1 \nu} \tag{3.5.2}$$

where

$$\nu = \nu_+ + \nu_- \tag{3.5.3}$$

and ρ is the density of the solution. M_1 and M_2 are the molecular weights of solvent and solute, respectively. At very low concentrations of electrolyte, (3.5.1) and (3.5.2) become

$$x_{+0} = 0.001 C_0 M_1 \nu_+ / \rho_0 \tag{3.5.4}$$
$$= 0.001 m_0 M_1 \nu_+ \tag{3.5.5}$$

where ρ_0 is the density of the solvent. Since the ratio a_+/a_{+0}, where a_{+0} is the activity a_+ of the cation at infinite dilution, is independent of the chosen standard state, and since the activity coefficients are unity at infinite dilution, we have

$$a_+/a_{+0} = f_+ x_+ / x_{+0} = y_+ C / C_{+0} = \gamma_+ m / m_{+0} \tag{3.5.6}$$

where f_+, y_+, and γ_+ are the activity coefficients on the mole fraction, molar, and molality scales respectively. Substituting (3.5.1), (3.5.2), (3.5.4), and (3.5.5) into (3.5.6) yields

$$f_+ = y_+(\rho - 0.001 C M_2 + 0.001 C M_1 \nu)/\rho_0 \tag{3.5.7}$$
$$f_+ = \gamma_+(1 + 0.001 m M_1 \nu) \tag{3.5.8}$$

Analogous equations exist for the anion. Since

$$f_\pm^\nu = f_+^{\nu_+} f_-^{\nu_-}; \quad \gamma_\pm^\nu = \gamma_+^{\nu_+} \gamma_-^{\nu_-}; \quad y_\pm^\nu = y_+^{\nu_+} y_-^{\nu_-} \tag{3.5.9}$$

then

$$f_\pm = y_\pm(\rho - 0.001 C M_2 + 0.001 C M_1 \nu)/\rho_0 \tag{3.5.10}$$

and

$$f_\pm = \gamma_\pm(1 + 0.001 m M_1 \nu) \tag{3.5.11}$$

Substituting numerical values in (3.5.10) yields

$$f_\pm = 1.55[1.246 + 0.0032(3 \times 18 - 111)]/1.0$$
$$= 1.65$$

From (3.5.10), (3.5.11), and

$$m = 1000 C/(1000 \rho - C M_2) \tag{3.5.12}$$

we can relate the molal to the molar activity coefficients and

$$\gamma_\pm = y_\pm(\rho - 0.001CM_2)/\rho_0 \tag{3.5.13}$$

and substituting numerical values into (3.5.13) yields

$$\gamma_\pm = 1.55[1.246 - (0.0032 \times 111)]/1.0$$

$$= 1.38$$

$\boxed{7}$ The ionic association constant K for the process

$$\text{Na}^+ + \text{Cl}^- \rightleftarrows \text{NaCl} \tag{3.7.1}$$

is given by

$$K = \theta(1 - \theta)^{-2}\, C^{-1} \tag{3.7.2}$$

where θ is fraction of associated ions and C is the total concentration of associated and unassociated Na^+ or Cl^-. The Fuoss treatment yields the relation

$$\theta = [4\pi(1 - \theta)^2\, CN_A a^3/3000]\, \exp[b/(1 + \kappa a)] \tag{3.7.3}$$

where

$$b = |\, z_+ z_- \,|\, e_0^2/a\epsilon kT \tag{3.7.4}$$

and κ, the Debye reciprocal length, is given by

$$\kappa = \left[4\pi \left(\sum_i n_i z_i^2 e_0^2\right)\Big/\epsilon kT\right]^{1/2} \tag{3.7.5}$$

where n_i is the concentration of ionic species of charge z_i in ions cm^{-3}. The distance of closest approach a is not simply the sum of the crystallographic radii and is best estimated semiempirically. Since the extended Debye–Hückel equation best fits the experimental data with $a = 4$ Å in the case of NaCl solutions (Falkenhagen and Ebeling, 1971) this value will be used here. Here, ϵ is the dielectric constant of the solvent.

Substituting (3.7.3) into (3.7.2) yields

$$K = (4\pi N_A a^3/3000)\, \exp[b/(1 + \kappa a)] \tag{3.7.6}$$

Evaluating b and κ from (3.7.4) and (3.7.5), respectively, yields, for aqueous $5\,M$ NaCl solution,

$$b = 1.79, \qquad \kappa = 7.36 \times 10^7 \text{ cm}^{-1}$$

Substituting numerical values into (3.7.6) yields

$$K = \frac{4\pi \times 6.023 \times 10^{23} \times (4 \times 10^{-8})^3}{3000} \exp \frac{1.79}{3.92}$$
$$= 0.254 \text{ liter mole}^{-1}$$

The units of the association constant arise from the units of C in Eq. (3.7.2).

For $10^{-6}\,M$ solution of NaCl in benzene, the following values are calculated in the same way:

$$b = 61.7, \qquad \kappa = 1.93 \times 10^5 \text{ cm}^{-1}$$

which give $K = 2 \times 10^{25}$ liters mole^{-1}.

| 9 | Equation (3.9.2) can be rewritten in the form |

$$K' = x/(a - x)(b - x) \tag{3.9.3}$$

where a and b are the total concentrations of Cu^{++} and SO_4^{--}, respectively, and x is the concentration of associated $CuSO_4$. Assuming the Beer–Lambert law, we can write

$$D_1 = xl\epsilon_{CuSO_4} + (a - x)\, l\epsilon_{Cu^{++}} \tag{3.9.4}$$

$$D_2 = al\epsilon_{Cu^{++}} \tag{3.9.5}$$

and subtracting (3.9.5) from (3.9.4) yields (3.9.1), which may be more conveniently written as

$$\Delta D = xl\, \Delta\epsilon \tag{3.9.6}$$

where ΔD and $\Delta\epsilon$ are written for $D_1 - D_2$ and $\epsilon_{CuSO_4} - \epsilon_{Cu^{++}}$, respectively. Substituting (3.9.6) into (3.9.3) yields

$$K' = \frac{\Delta D/(l\, \Delta\epsilon)}{[a - (\Delta D/l\, \Delta\epsilon)][b - (\Delta D/l\, \Delta\epsilon)]} \tag{3.9.7}$$

which rearranges to

$$\frac{ab}{\varDelta D} = \frac{a + b - x}{l\,\varDelta\epsilon} + \frac{1}{K'l\,\varDelta\epsilon} \qquad (3.9.8)$$

and K' can be found from the slope and intercept of a plot of $ab/\varDelta D$ against $(a + b - x)$. Since x is unknown, the slope and intercept must be found by an iterative procedure. An initial plot $ab/\varDelta D$ is made against $(a + b)$, i.e., x is initially assumed to be zero. From the slope of this initial plot, S_{int}, approximate values of x are calculated. Since the slope of the initial plot is approximately $(l\,\varDelta\epsilon)^{-1}$, we have, from (3.9.6),

$$x_{approx} = (\varDelta D)\,S_{int} \qquad (3.9.9)$$

A more precise estimate of the slope and intercept can be obtained by plotting $ab/\varDelta D$ against $(a + b - x_{approx})$ and the procedure repeated until a constant slope and intercept are obtained. In the present case, the first plot of $ab/\varDelta D$ against $(a + b - x_{approx})$ gives satisfactory results, and details of the calculation are shown in Table 3.9.2.

The value of K' calculated from the first plot of $ab/\varDelta D$ against $(a + b - x_{approx})$ is 18.1 liters mole^{-1}, which is close to the final computer-calculated value of 18.9 liters mole^{-1}. Values of K' together with the average ionic strength of the solutions are given in Table 3.9.3.

TABLE 3.9.2.   Details of Calculation of K' for CuSO$_4$
in Aqueous Solution at Ionic Strength I_1

$\varDelta D$	$10^3(a + b)$, M	$10^6 ab/\varDelta D$	$10^3 x_{approx}$,* M	$10^3 x$,† M
0.143	12.05	227	0.47	0.52
0.206	16.05	236	0.68	0.75
0.260	20.05	249	0.86	0.95
0.312	24.05	260	1.03	1.13
0.352	28.05	276	1.16	1.28
0.389	32.05	292	1.28	1.42

* Calculated from an initial slope of 3.3×10^{-3}.
† Computer solution.

TABLE 3.9.3.   Values of K′ at Different Ionic Strengths*

I	$\bar{I}$	K' liters mole^{-1}
I_1	0.090	18.9
I_2	0.070	28.4
I_3	0.058	31.5

* $\bar{I}$ is the average ionic strength.

The ionic strength of each solution is given by

$$I = \tfrac{1}{2} \sum_i C_i z_i^2 \tag{3.9.10}$$

$$= 3a + 3b + C_{\text{NaClO}_4} - 4x \tag{3.9.11}$$

The ionic association constant K is related to K' by

$$K = K'y/y_{\pm}^2 \tag{3.9.12}$$

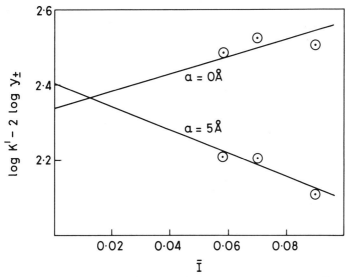

Fig. 3.9.1. Estimation of the association constant of copper sulfate by extrapolation of values of log K' − 2 log $y_{\pm}$ to infinite dilution.

where y is the molar activity coefficient of undissociated $CuSO_4$ and $y_\pm$ is the molar mean ion activity coefficient of dissociated $CuSO_4$,

$$\log K = \log K' - 2 \log y_\pm \qquad (3.9.13)$$

where $\log y$ has been neglected since the activity coefficient of the uncharged species, $CuSO_4$, will be close to unity.

To obtain an estimation of K, $y_\pm$ can be calculated from the Debye–Hückel law:

$$\log y_\pm = - \mid z_+ z_- \mid AI^{1/2}/(1 + aBI^{1/2}) \qquad (3.9.14)$$

However, there is a problem in deciding what value to use for the distance of closest approach, a, in a mixture of electrolytes. The procedure adopted here has been to select two extreme values of a (0 and 5 Å), plot $\log K' - 2 \log y_\pm$ against I, and extrapolate both plots to infinite dilution (Fig. 3.9.1). The two values of K so obtained may be considered to be the extremes of the range of values of K indicated by the data. Figure 3.9.1 shows that K lies in the range 210–260 liters mole^{-1}.

The most appropriate value of a would yield a plot of $\log K' - 2 \log y_\pm$ against I that is parallel to the I axis. When sufficient data are available, this value of a can be found by an iterative procedure and K obtained.

CHAPTER 4

ION TRANSPORT IN SOLUTIONS

1 Calculate the conventional and the absolute Stokes mobilities of a univalent ion of radius 2.0 Å in a solution of viscosity 0.01 P.

2 According to Faraday's law, passage of a current I for time t leads to the deposition or dissolution of a number of ions $N_A It/nF$, where n is the number of electrons involved per ion and N_A is Avogadro's number.

However, in ionic solutions, each ion carries a different fraction of current. Explain the apparent paradox between Faraday's law and the concept of transport numbers. Formulate the essence of your explanation in an equation.

3 What is the root mean square distance traversed by chloride ions in aqueous solution in 1 min?

Chlorine is produced in a $10^{-3} M$ solution of chloride ions, containing a high concentration of indifferent electrolyte, by application of a current step of 1 mA cm^{-2}. Calculate the concentration of chloride ions 50 μm from the surface of the electrode 1 sec after the current is switched on. $D_{Cl^-} = 2.03 \times 10^{-5}$ cm^2 sec^{-1}.

4 State the differences among Einstein's equation, the Stokes–Einstein equation, the Nernst–Einstein equation, and the Nernst–Planck equation. Why does the factor of 300 appear in the latter two equations? If the diffusion coefficient of a divalent ion is

$5 \times 10^{-5} \, \text{cm}^2 \, \text{sec}^{-1}$, calculate its mobility. To what degree do you think your answer will be correct?

$\boxed{5}$ Calculate Walden's product for a univalent ion of radius 3 Å at 25° C for water. At infinite dilution, what would be the equivalent conductivity of the ion in methanol? $\eta_{CH_3OH} = 5.46 \times 10^{-3}$ P.

$\boxed{6}$ Calculate the concentration gradient established at an electrode solution interface when a univalent ion migrates toward the electrode in a solution of concentration 1 M at 25°C. The field in the diffusion layer to which the ion is subject is 10^5 V cm^{-1}.

$\boxed{7}$ The potassium chloride salt bridge is often used to minimize liquid–liquid junction potentials. Calculate the liquid–liquid junction potential when saturated KCl solution is in contact with (a) 0.01 M HCl, (b) 5 M CaCl$_2$ solution using the data of Table 4.7.1.

TABLE 4.7.1.
Transport Numbers of Cations

Electrolyte	t_+
KCl	0.491
HCl	0.821
CaCl$_2$	0.438

$\boxed{8}$ The relaxation effect on the asymmetry of ionic clouds in solution is related to the time it takes for the atmosphere to form. Show that this is given by

$$1/2D\kappa^2$$

where κ^{-1} is the thickness of the ionic atmosphere.

At what frequency of an electric field applied to 0.001 N LiCl solution would the ionic atmosphere relaxation effect become undetectable?

$$D_{LiCl} = 1.27 \times 10^{-5} \, \text{cm}^2 \, \text{sec}^{-1}$$

9   Some physical properties of 1-propanol–water mixtures are given in Table 4.9.1 and the equivalent conductivity of sodium chloride at various concentrations in 1-propanol–water mixtures is given in Table 4.9.2. Using the Fuoss approach, estimate the association constant of sodium chloride in each solvent mixture (Goffredi and Shedlovsky, 1967).

TABLE 4.9.1.   Some Physical Properties
of 1-Propanol–Water Mixtures at 35°C

Wt. % C_3H_7OH	Dielectric constant	10^3 Viscosity, P	Density, g ml^{-1}
60.0	34.49	19.40	0.8766
80.0	24.70	18.02	0.8350
90.0	21.34	16.62	0.8140

TABLE 4.9.2.   Equivalent Conductivities and Mean Ion Activity
Coefficients of NaCl in 1-Propanol–Water Mixtures at 35°C

Wt. % C_3H_7OH	$10^4\ C_{NaCl}$, M	Λ, mho cm^2 equiv^{-1}	$y_\pm$
60.0	5.0869	45.337	2.63
	11.130	44.269	2.25
	18.452	43.352	2.03
	24.239	42.759	1.92
	33.823	41.959	1.79
80.0	4.448	35.081	2.08
	10.986	33.205	1.80
	17.716	31.865	1.67
	24.607	30.812	1.59
	34.212	29.652	1.51
	42.774	28.809	1.46
90.0	1.4379	33.229	1.73
	4.0318	31.433	1.48
	6.6842	30.150	1.38
	8.6972	29.379	1.33
	11.524	28.465	1.28
	14.322	27.702	1.24

[10] Ions migrating in high applied electric fields can outrun their ionic atmospheres, i.e., in the relaxation time of the atmosphere, the ions migrate further than κ^{-1}. This is known as the Wien effect (Bockris and Reddy, 1970) and causes the conductivity of the electrolyte solution to increase with increasing field. Calculate the field strength required to bring about a 5% deviation from Ohm's law by the Wien effect for 0.01 N KCl solution at 25°C. $\Lambda^\circ_{\text{KCl}} = 149.8$ ohm^{-1} cm^2 equiv^{-1}; $\eta_w = 8.90 \times 10^{-3}$ P; $\epsilon_w = 78.3$; $D_{\text{KCl}} = 1.92 \times 10^{-5}$ cm^2 sec^{-1}.

Deviations from Ohm's law can also result at high fields since it is then no longer a good approximation to relate the drift velocity linearly to the field strength. For the above solution, what field strength is required to bring about a 5% deviation from Ohm's law by this effect?

[11] What, in general terms, is the primitive of Fick's second law? Why are three initial and boundary conditions needed for the solution of this differential equation?

Use Laplace transformations to deduce the concentration response in the solution adjacent to soluble metal electrode when a constant anodic current is switched on at $t = 0$. Correspondingly, show that, if a sinusoidal current is applied about the reversible potential, then

$$C(x = 0) = C_0 - \frac{I_{\text{max}}}{AnF} (D\omega)^{-1/2} \cos\left(\omega t - \frac{\pi}{4}\right) \quad (4.11.1)$$

where C_0 is the concentration of the metal ions of charge n in the bulk of the solution, I_{max} is the amplitude of the current wave of frequency $\omega/2\pi$, A is the area of the electrode, and D is the diffusion coefficient of the metal ions in aqueous solution.

[12] Radiotracer techniques have been used to measure the diffusion coefficients of Cl$^-$ and Na$^+$ ions in fused sodium chloride. These values are given by

$$D_{\text{Cl}^-} = 1.9 \times 10^{-3} \exp(-7400/RT) \text{ cm}^2 \text{ sec}^{-1}$$

$$D_{\text{Na}^+} = 2.1 \times 10^{-3} \exp(-7100/RT) \text{ cm}^2 \text{ sec}^{-1}$$

Calculate the equivalent conductivity of fused sodium chloride at

935°C using the Nernst–Einstein equation and compare the result with the experimental value of 156.4.

Discuss this discrepancy in terms of the Onsager reciprocity relations. The Nernst–Einstein equation is a phenomenological relation. *Can* it be incorrect? Suggest structural hypotheses for its inapplicability.

ANSWERS

1 The absolute Stokes mobility of an ion $\bar{u}_{ab}$ is given by

$$\bar{u}_{ab} = (6\pi r \eta)^{-1} \qquad (4.1.1)$$

where r is the radius of the ion and η the viscosity of the solvent. Substituting numerical values into (4.1.1) yields

$$\bar{u}_{ab} = (6\pi \times 2 \times 10^{-8} \times 0.01)^{-1}$$
$$= 2.7 \times 10^8 \text{ cm sec}^{-1} \text{ dyne}^{-1}$$

The conventional mobility u_{conv} is given by

$$u_{\text{conv}} = ze_0/1800\pi r \eta$$

$$= \frac{4.8 \times 10^{-10}}{1800\pi \times 2 \times 10^{-8} \times 0.01}$$

$$= 4.31 \times 10^{-4} \text{ cm}^2 \text{ sec}^{-1} \text{ V}^{-1} \qquad (4.1.2)$$

3 The root mean square distance x_{rms} traveled by an ion executing random walk motions in time t is given by

$$x_{\text{rms}} = (2Dt)^{1/2} \qquad (4.3.1)$$

where D is the diffusion coefficient. Substituting given values into (4.3.1) gives

$$x_{\text{rms}} = (2 \times 2.03 \times 10^{-5} \times 60)^{1/2}$$
$$= 4.9 \times 10^{-2} \text{ cm}$$

The concentration of chloride ions C at any time t and distance from the electrode x is given by the solution of Fick's second law. The initial condition

$$C[t = 0] = C^0 \qquad (4.3.2)$$

and the boundary conditions:

$$C[x \rightarrow \infty] = C^0 \qquad (4.3.3)$$

and

$$(\partial C / \partial x)_{x=0} = -J/D \qquad (4.3.4)$$

must be satisfied. Condition (4.3.2) implies that, before the current step is switched on, the concentration of chloride ions is uniform and everywhere equal to that in the bulk of the solution, C^0. The implication of condition (4.3.3) is that, far from the electrode, the concentration of chloride ions remains unchanged, while the implication of condition (4.3.4) is that the flux J of chloride ions at the electrode is described by Fick's first law. The solution of Fick's second law, under these conditions, is given by

$$C = C^0 - \frac{J}{D^{1/2}} \left[\frac{2t^{1/2}}{\pi^{1/2}} \exp\left(-\frac{x^2}{4Dt}\right) - \frac{x}{D^{1/2}} \operatorname{erfc}\left(\frac{x^2}{4Dt}\right)^{1/2} \right] \qquad (4.3.5)$$

where the flux is given by

$$J = i/F \text{ mole cm}^{-2} \text{ sec}^{-1}$$

and i is the current density of the step. A consideration of units in (4.3.5) shows that the concentration terms must be expressed in moles cm^{-3}. Substituting numerical values into (4.3.5) yields

$$C = 10^{-6} - \frac{10^{-3}}{96,500(2.03 \times 10^{-5})^{1/2}} \left[\frac{2}{\pi^{1/2}} \exp\left(-\frac{(5 \times 10^{-3})^2}{4 \times 2.03 \times 10^5 \times 1}\right) \right.$$

$$\left. - \frac{5 \times 10^{-3}}{(2.03 \times 10^{-5})^{1/2}} \operatorname{erfc}\left(\frac{(5 \times 10^{-3})^2}{4 \times 2.03 \times 10^5 \times 1}\right)^{1/2} \right]$$

$$= 10^{-6} - 2.3 \times 10^{-6}[0.830 - 1.11 \operatorname{erfc}(0.554)] \qquad (4.3.6)$$

Since

$$\operatorname{erfc}(X) = 1 - \operatorname{erf}(X) \qquad (4.3.7)$$

by finding the error function erf(0.554) from tables, we arrive at

$$C = 10^{-6} - 2.3 \times 10^{-6}(0.830 - 1.11 \times 0.435)$$
$$= 2.0 \times 10^{-7} \text{ mole cm}^{-3}$$
$$= 2.0 \times 10^{-4} \text{ mole liter}^{-1}$$

5 Walden's product $\Lambda\eta$, where Λ is the equivalent conductance of the ionic species and η the viscosity of the solvent, can be approached theoretically by eliminating the diffusion coefficient D from the Nernst–Einstein equation,

$$\Lambda = (zF^2/RT)D \qquad (4.5.1)$$

and the Stokes–Einstein equation,

$$D = kT/6\pi r\eta \qquad (4.5.2)$$

The result is

$$\Lambda\eta = ze_0F/6\pi r \qquad (4.5.3)$$

Substituting numerical values into (4.5.3) gives

$$\Lambda\eta = \frac{4.8 \times 10^{-10} \times 96,500}{6\pi \times 3 \times 10^{-8}}$$
$$= 82 \text{ esu coulomb cm}^{-1}\text{ equiv}^{-1} \qquad (4.5.4)$$

The conventional units of Walden's product are mho cm^2 equiv^{-1} P. One way of converting the units is by using

$$\frac{\text{esu coulomb}}{\text{cm equiv}} = \frac{\text{esu sec V mho}}{\text{cm equiv}}$$
$$= \frac{0.333 \times 10^{-2}(\text{esu})^2 \text{ sec mho}}{\text{cm}^2 \text{ equiv}}$$
$$= 0.333 \times 10^{-2} \text{ dyne sec mho equiv}^{-1}$$
$$= 0.333 \times 10^{-2} \text{ mho cm}^2 \text{ equiv}^{-1} \text{ poise}$$

Consequently, Walden's product in conventional units is obtained by multiplying (4.5.4) by 0.333×10^{-2},

$$\Lambda\eta = 0.272 \text{ mho cm}^2 \text{ equiv}^{-1} \text{ poise}$$

In methanol, the individual ionic equivalent conductivity is found from Walden's product to be

$$\Lambda = 0.272/5.46 \times 10^{-3} \text{ mho cm}^2 \text{ equiv}^{-1}$$
$$= 49.8 \text{ mho cm}^2 \text{ equiv}^{-1}$$

$\boxed{7}$ The potential difference between two solutions in contact is the diffusion potential resulting from the diffusion of all the ions in the two solutions. For two solutions designated I and II, we have from Planck (1890a,b) (cf. Bockris and Reddy, 1970):

$$\psi_I - \psi_{II} = \frac{RT}{F} \sum_i \int_I^{II} \frac{t_i}{z_i} d(\ln a_i) \tag{4.7.1}$$

The transport number of the ith species, t_i, is defined in terms of the concentrations C_i (in equivalents per unit volume) and conventional mobilities u_i of the species in the two solutions by

$$t_i = C_i u_i FX \Big/ \sum_i C_i u_i FX \tag{4.7.2}$$

where X is the electric field. Since X is constant, (4.7.2) becomes

$$t_i = C_i u_i \Big/ \sum_i C_i u_i \tag{4.7.3}$$

Assuming ideal solutions, we can write $d(\ln C_i) = dC_i/C_i$ for $d(\ln a_i)$ in (4.7.1); substituting (4.7.3) for t_i yields

$$\psi_I - \psi_{II} = \frac{RT}{F} \sum_i \int_I^{II} \frac{u_i/z_i}{\sum_i C_i u_i} dC_i \tag{4.7.4}$$

Making the assumption that the concentration of the ith species at some point in the diffusion layer, $C_i(x)$, is given by

$$C_i(x) = C_{i,I} + (C_{i,II} - C_{i,I}) x$$

where x is a positive fraction, (4.7.4) becomes

$$\psi_I - \psi_{II} = \frac{RT}{F} \sum_i \int_0^1 \frac{(u_i/z_i)(C_{i,II} - C_{i,I}) dx}{\sum_i C_{i,I} u_i + x \sum_i (C_{i,II} - C_{i,I}) u_i} \tag{4.7.5}$$

Evaluating the integral of (4.7.5) yields

$$\psi_{\mathrm{I}} - \psi_{\mathrm{II}} = \frac{RT}{F} \frac{\sum_i (C_{i,\mathrm{II}} - C_{i,\mathrm{I}}) u_i/z_i}{\sum_i (C_{i,\mathrm{II}} - C_{i,\mathrm{I}}) u_i} \ln \frac{\sum_i C_{i,\mathrm{II}} u_i}{\sum_i C_{i,\mathrm{I}} u_i} \qquad (4.7.6)$$

and (4.7.6) is known as the Henderson equation (Henderson, 1907, 1908).

For a solution of pure potassium chloride, we have, from (4.7.3),

$$t_{\mathrm{K}^+} = u_{\mathrm{K}^+}/(u_{\mathrm{K}^+} + u_{\mathrm{Cl}^-}) \qquad (4.7.7)$$

and rearranging yields

$$u_{\mathrm{Cl}^-} = u_{\mathrm{K}^+}(1 - t_{\mathrm{K}^+})/t_{\mathrm{K}^+} \qquad (4.7.8)$$

and substituting for t_{K^+} the value of 0.491 (Table 4.7.1) into (4.7.8) gives

$$u_{\mathrm{Cl}^-} = 1.037 u_{\mathrm{K}^+} \qquad (4.7.9)$$

Similarly, we have, making use of the data of Table 4.7.1,

$$u_{\mathrm{H}^+} = 4.760 u_{\mathrm{K}^+} \qquad (4.7.10)$$

$$u_{\mathrm{Ca}^{2+}} = 0.808 u_{\mathrm{K}^+} \qquad (4.7.11)$$

Consider now the liquid–liquid junction

$$\mathrm{I} \qquad\qquad \mathrm{II}$$
$$\mathrm{KCl(sat)} \| \mathrm{HCl}(0.01\ M)$$

The concentration of saturated KCl solution at room temperature is 4.2 M; substituting the values for concentrations and ionic mobilities given by Eqs. (4.7.9) and (4.7.10) in (4.7.6) yields

$$\psi_{\mathrm{I}} - \psi_{\mathrm{II}} = \frac{RT}{F} \left[\frac{0.01(4.760 - 1.037) - 4.2(1 - 1.037)}{0.01(4.760 + 1.037) - 4.2(1 + 1.037)} \right.$$

$$\left. \times \ln \frac{0.01(4.760 + 1.037)}{4.2(1 + 1.037)} \right]$$

$$= 0.059(0.197/-8.502) \log(0.058/8.56)$$

$$= 0.0029\ \mathrm{V}$$

Similarly, for the liquid–liquid junction

$$\text{I} \qquad \text{II}$$
$$\text{KCl(sat)} \| \text{CaCl}_2(5\ M)$$

we have, by substituting (4.7.9) and (4.7.11) into (4.7.6),

$$\psi_\text{I} - \psi_\text{II} = 0.059(-6.18/9.89) \log(18.45/8.56)$$
$$= 0.012\ \text{V}$$

| 9 | For the association of sodium chloride

$$\text{Na}^+ + \text{Cl}^- \rightleftarrows \text{NaCl} \tag{4.9.1}$$

the association constant K_A is defined by

$$K_A = C_\text{NaCl}/C_{\text{Na}^+}C_{\text{Cl}^-} \tag{4.9.2}$$

where C_NaCl, C_{Na^+}, and C_{Cl^-} are the actual concentrations of the species NaCl, Na$^+$, and Cl$^-$, respectively, existing in the solution.

The Fuoss approach to the estimation of association constants from conductivity data leads to the relation

$$\frac{Z}{\Lambda} = \frac{1}{\Lambda^\circ} + \frac{K_A \Lambda C y_\pm^2}{(\Lambda^\circ)^2 Z} \tag{4.9.3}$$

where Λ^0 is the equivalent conductivity at infinite dilution, Λ that at concentration C, and $y_\pm$ the mean ion activity coefficient on the molar scale. The function Z is given by

$$Z = \tfrac{4}{3} \cos^2\{\tfrac{1}{3} \cos^{-1}[-\tfrac{1}{2}(3^{3/2}z^*)]\} \tag{4.9.4}$$

where

$$z^* = (A + B\Lambda^\circ)(C\Lambda)^{1/2}/(\Lambda^\circ)^{3/2} \tag{4.9.5}$$

and

$$A = (ze_0F/900\pi\eta)(8\pi z^2e_0^2N_A/1000\epsilon kT)^{1/2} \tag{4.9.6}$$
$$B = (e_0^2w/6\epsilon kT)(8\pi z^2e_0^2N_A/1000\epsilon kT)^{1/2} \tag{4.9.7}$$

where η is the viscosity and, for a 1:1 electrolyte, $w = 0.5859$.

From Eq. (4.9.3), it is seen that the association constant can be found from the slope of the plot of Z/Λ against $Cy_{\pm}^2/Z(\Lambda°)^2$. However, it is first necessary to evaluate $\Lambda°$ and Z. Recalling Kohlrausch's law

$$\Lambda = \Lambda° - A^*C^{1/2} \qquad (4.9.8)$$

where A^* is a positive constant, it can be seen that $\Lambda°$ can be obtained from the extrapolation of a plot of Λ against $C^{1/2}$ to infinite dilution. The procedure does not yield highly accurate values of $\Lambda°$ since the degree of association changes with concentration and A^* is not strictly a constant under these circumstances. Figure 4.9.1 shows a plot of Λ against $C^{1/2}$ for the data of Table 4.9.2. The equivalent conductances at infinite dilution at 60, 80, and 90% C_3H_7OH are 47.4, 38.0, and 35.4 mho cm^2 equiv^{-1}, respectively.

The term Z is evaluated making use of Eqs. (4.9.4)–(4.9.7). Values of A and B for various solvent compositions are listed in Table 4.9.3. Values of Z and intermediate values in the calculation of Z are listed for various salt concentrations in 60% C_3H_7OH in Table 4.9.4.

Table 4.9.5 lists values of Z/Λ and $\Lambda Cy_{\pm}^2/Z(\Lambda°)^2$ for 60% C_3H_7OH

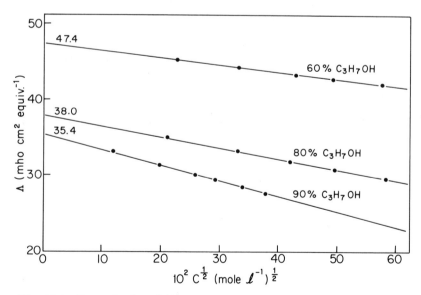

Fig. 4.9.1. Determination of Λ^0 by extrapolating plots of Λ against $C^{1/2}$ to infinite dilution for sodium chloride in 60, 80, and 90% C_3H_7OH solutions.

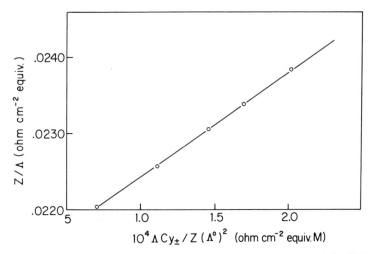

Fig. 4.9.2.   Plot of Z/Λ against $\Lambda Cy_{\pm}^{2}/Z(\Lambda^{0})^{2}$ for sodium chloride in 60% C_3H_7OH at 35°C.

TABLE 4.9.3.   Values of A and B for 1-Propanol–Water Mixtures

Wt. % 1-propanol	A	B
60.0	41.2	0.75
80.0	52.5	1.24
90.0	61.2	1.56

TABLE 4.9.4.   Calculations of Z for Various Salt Concentrations in 60% 1-Propanol–Water Mixture at 35°C

$10^4 C_{NaCl}$, M	$10z^*$	$\frac{1}{3}\cos^{-1}[\frac{1}{2}(3^{3/2}z^*)]$	Z
5.089	11.28	29° 57′	0.9986
11.130	16.45	29° 55′	0.9993
18.452	20.92	29° 54′	0.9995
24.239	23.97	29° 53′	0.9998
33.823	27.97	29° 52′	1.0002

TABLE 4.9.5.   Values of Z/Λ and $\Lambda Cy_{\pm}^2/(\Lambda°)^2 Z$
for 60% C_3H_7OH at 35°C

Z/Λ, ohm cm^{-2} equiv	$\Lambda Cy_{\pm}^2/Z(\Lambda^0)^2$, ohm cm^{-2} equiv M
0.02203	0.711×10^{-4}
0.02257	1.11×10^{-4}
0.02306	1.46×10^{-4}
0.02338	1.70×10^{-4}
0.02384	2.02×10^{-4}

at 35°C and these values are plotted in Fig. 4.9.2. From the slope of this plot, the association constant K_A is found to 13.4 liters mole^{-1}.

A more precise method of finding K_A from the given data is the numerical solution of (4.9.3). Values of $\Lambda°$ are assumed and values of K_A calculated at each concentration until the value of $\Lambda°$ which yields a value of K_A independent of concentration is obtained. Table 4.9.6 lists computer calculations of $\Lambda°$ and K_A obtained in this way (Goffredi and Shedlovsky, 1967) for 60, 80, and 90% C_3H_7OH. It can be seen that there is close agreement between our values of $\Lambda°$ and K_A and the values obtained by the numerical method. However, this would probably not be the case for less precise data.

TABLE 4.9.6.   Association Constants and Values of $\Lambda°$
for NaCl in 1-Propanol–Water Mixtures at 35°C

Wt. % C_3H_7OH	Λ^0, mho cm^2 equiv^{-1}	K_A, liters mole^{-1}
60	47.38	13 ± 1
80	37.96	45 ± 1
90	35.34	158 ± 3

CHAPTER 5

PROTONS IN SOLUTION

<table>
<tr><td>**1**</td><td>$H_9O_4^+$ is often given as a reasonable configuration for the proton in solution. Using the Born equation and the equations</td></tr>
</table>

for ion–dipole and ion–quadrupole interactions, show that the proton affinity of water must be about 180 kcal mole^{-1}. The absolute individual heat of hydration of protons in water is -263 kcal mole^{-1}.

$$\epsilon_w = 78.5; \qquad d\epsilon_w/dT = -0.356 \text{ deg}^{-1}$$

$$\mu_w = 1.87 \times 10^{-18} \text{ esu cm}; \qquad r_{H_3O^+} = 1.05 \text{ Å}$$

$$r_w = 1.38 \text{ Å}; \qquad p_w = 3.9 \times 10^{-26} \text{ esu cm}^2$$

<table>
<tr><td>**2**</td><td>Using the data of Table 5.2.1, deduce the percentage Grotthuss conduction, or anomalous proton conduction, of each solvent.</td></tr>
</table>

TABLE 5.2.1. Equivalent Conductance at Infinite Dilution of Hydrogen and Lithium Chlorides in Water, Methanol, Ethanol, and Propanol at 20°C

Property investigated	Solvent			
	Water	Methanol	Ethanol	n-Propanol
Λ of hydrogen chloride	426.2	192	84.3	22
Λ of lithium chloride	115.0	90.9	38.0	18

<table>
<tr><td>**3**</td><td>The proton affinity of water is -182 kcal mole^{-1}. Using the data of Table 5.3.1, calculate the deuteron affinity of water.</td></tr>
</table>

45

TABLE 5.3.1.
Zero-Point Vibrational Frequencies
(cm^{-1}) for H_3O^+ and DH_2O^+ *

H_3O^+	DH_2O^+
3500	3500
3500	3500
3500	2600
1520	1520
1520	1320
1520	1320

* Data from Conway (1958).

---

4 (a) The equivalent conductance of HCl in methanol–water mixtures plotted against the mole per cent methanol gives a minimum at a certain concentration of methanol. Why?

(b) Why is anomalous proton mobility most pronounced in water?

(c) Give examples of amphiprotic solvents.

(d) Define proton free energy levels.

(e) Explain with examples: "primary solvent effect on acid strength."

---

5 From classical mechanics, show that the frequency of free rotation of a water molecule is about 10^{13} sec^{-1}. However, the dielectric relaxation time is 0.55×10^{-11} sec^{-1}, with an activation energy of 3.8 kcal $mole^{-1}$ at 300°K. Interpret these two results qualitatively. The moment of inertia of a water molecule is 2×10^{-40} g cm^2.

Show that the rate of proton transfer per water molecule is about $5 M \times 10^{11}$ sec^{-1}, where M is the molarity of protons in the solution. The actual rate of proton exchange transfer in either direction has been shown to be 7.5×10^5 cm sec^{-1} (Conway et al., 1956). Take the radius of a water molecule as 1.4 Å.

The rotation of a water molecule in solution requires the breaking and reformation of a number of hydrogen bonds to neighboring water molecules in the water structure. An approximate value of the minimum activation energy for this process may be taken as that estimated for the rotation of a single water molecule, 10 kcal mole^{-1}, which neglects the cooperative rotation of neighboring water molecules (Conway *et al.*, 1956). Calculate the maximum rate of thermal reorientation of water molecules. What considerations lead you to a decision as to whether this breaking is rate-determining in aqueous solutions? Calculate your decision.

---

6   In solution, the vibrational levels in H_3O^+ are changed by the interaction of H_3O^+ with the solvent. Several IR and Raman studies have been done on aqueous solutions of strong acids, but assignment of frequencies from such work is difficult because the hydroxonium ion bands appear as broad shoulders on the H_2O bands. O'Ferrall *et al.* (1971) have made a detailed vibrational analysis of the hydroxonium ion in solutions and found that their vibrational frequencies agreed with the values experimentally obtained for the crystal of $H_9O_4{}^+Br^-$ but not with the experimental values obtained for the hydroxonium halides salts in solutions of sulfur dioxide. This seems to indicate that the hydroxonium ion in solutions exist more like an $H_9O_4{}^+$ than H_3O^+.

TABLE 5.6.1. Calculated and Observed Frequencies* for Solutions of H_3O^+ Salts in SO_2 and Crystalline $H_9O_4{}^+Br^-$

	H_3O^+/SO_2			$H_9O_4{}^+Br^-$	
	IR^{obs}	Raman	Calc.	IR^{obs}	Calc.
$\nu_1(a)$	3405	3415	3410	2060	2640
$\nu_3(c)$	3470	3447	3455	2630	2060
$\nu_2(a)$	—	—	785	1313	1320
$\nu_4(c)$	1670	1700	1690	1845	1840
$\nu_L(a)$	—	—	—	902	870
(c)	—	—	—	738	920
$\nu_T(a)$	—	—	—	569	405
(c)	—	—	—	556	570
ν_b	—	—	—	78	97–27

* Frequencies in units of cm^{-1}. ν_1, ν_2, ν_3, and ν_4 are the bending and stretching frequencies, while ν_L, ν_T, and ν_b have been assigned to libration, hindered translation, and hydrogen-bond bending, respectively.

The wave numbers for the different vibrations are tabulated in Table 5.6.1. Levich (1970) asserts that a significant rate of transfer of electrons from an electrode to a proton in H_3O^+ cannot occur because (a) the difference in successive energy levels in H_3O^+ is so large that Tafel lines would not be smooth; (b) there will not be a sufficient occupancy of higher vibrational or rotational levels to maintain the observed rate. Calculate the energy interval (in meV) between rotational states in H_3O^+. Calculate the needed number of states to accommodate typical reaction rates (e.g., 10^{-10} A cm^{-2}, with a heat of activation of 20 kcal mole^{-1}). Thus, examine whether the thermal equilibrium of H_3O^+ with water provides a situation consistent with this mechanism of the activation.

$\boxed{7}$ The pK_w values for pure ice and pure supercooled water at $-10°C$ are 20.6 and 15.1, respectively, and the specific conductivities are practically the same with a value of 1.4×10^{-9} mho cm^{-1}. Calculate the proton mobility in ice and water at $-10°C$ taking the transport number of the proton in ice to be the same as that in water, 0.64.

Give a qualitative discussion of the effects of proton concentration on the mechanism of proton transfer. Using the results of Problem 5 and the information that the spontaneous thermal rotation rate of water molecules in ice is 10^6 sec^{-1}, show that the proton mobility in ice is consistent with proton tunneling as the rate-determining step for proton transfer in ice.

$\boxed{8}$ Assess the relative contributions of Wynne–Jones, Gurney, and Bell to the theory of the strength of acids.

$\boxed{9}$ Proton free energy levels (pfel) are shown in Fig. 5.9.1. Use these data to calculate the pK_A values for acetic and chloroacetic acids. What is the relative strength of chloroacetic acid with respect to acetic acid? Estimate the relative strength of chloroacetic acid with respect to β-chloropropionic acid in methanol. Take $\epsilon_w = 78.5$ and $\epsilon_{MeOH} = 32.5$ and the effective radii of the β-chloropropionic and chloroacetic ions as 4.2 Å and 2.9 Å, respectively. K_A for β-chloropropionic acid is 1.0×10^{-4}.

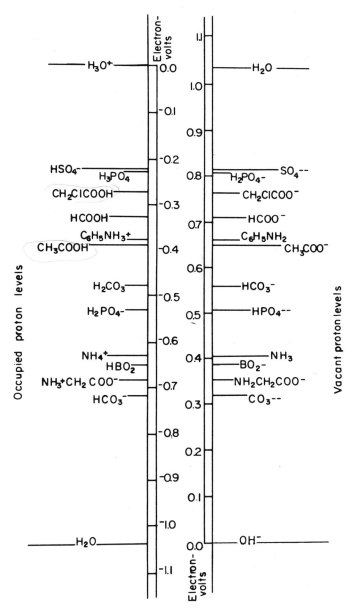

Fig. 5.9.1.   Proton free energy levels for a selection of acids in aqueous solution.

10 Proton transfer from an H_3O^+ to a suitably oriented H_2O in solution takes place by tunneling. Write the Schrödinger equation, in one dimension, appropriate to this situation and find the probability that a proton which presents itself at the barrier will tunnel. Assume an Eckhardt barrier and express the probability that the proton will tunnel at an energy W, in terms of the barrier width, the energy of activation, and the difference in the initial- and final-state energies. Express the initial- and final-state energies in terms of the applied field.

Show that for the low-field approximation and assuming a classical distribution of protons,

$$\vec{N} - \overleftarrow{N} = (N_0/kT)\left\{\int_0^\infty \vec{P}_w \exp[-(W - \phi_x)/kT]\, dw\right.$$

$$\left. - \int_0^\infty \overleftarrow{P}_w \exp[-(W + \phi_x)/kT]\, dw\right\} \qquad (5.10.1)$$

where $\vec{N}$ and $\overleftarrow{N}$ are the numbers of particles that can permeate the barrier in the forward and reverse directions, respectively, in unit time; N_0 is the number of particles approaching the barrier in unit time that may permeate it; ϕ_x is a function of the applied field; and

$$\vec{P}_w = \frac{\cosh 2KW^{1/2} - \cosh(\phi_x/W)\,KW^{1/2}}{\cosh 2KW^{1/2} - \cosh \pi\{[4(E^* - \phi_x) - C]/C\}^{1/2}}$$

$$\overleftarrow{P}_w = \frac{\cosh 2KW^{1/2} - \cosh(\phi_x/W)\,KW^{1/2}}{\cosh 2KW^{1/2} - \cosh \pi\{[4(E^* + \phi_x) - C]/C\}^{1/2}}$$

where $K = \pi l(2m)^{1/2}/h$, with m the mass of the proton and l the barrier half-width; ϕ is the electric potential energy in the activated state; $C = (h^2/8)\,ml^2$; W, the total energy, is a function of the position x across the barrier; E^* is the activation energy.

11 Use the Morse equation quantitatively to construct potential energy curves to represent the transfer of a proton from H_3O^+ to H_2O in solution. Take the radius of the hydrogen atom as 0.3 Å, the OH bond distance as 1.05 Å, and the effective radius of a water molecule in a hydrogen-bonded situation as 1.4 Å. The force constant of the OH bond is 5.31×10^{-5} dyne cm^{-1}, the dissociation energy of a proton from H_3O^+ is 180 kcal mole^{-1}, and the dissociation energy of an H_3O^+ in water is 263 kcal mole^{-1}. The resonance energy at the activated complex may be approximated as about half the hydrogen bond energy.

Make a numerical calculation of the fraction of protons that tunnel through the barrier using the considerations expressed in Problem 10 and assuming a classical distribution of protons.

Eigen and deMayer (1958) rightly suggested that a quantized distribution of proton energies should be used to calculate the fraction of protons tunneling through the barrier. By numerical calculation, show that the fraction of protons tunneling through the barrier calculated in this way is about twice that calculated using a classical distribution. Calculate the net rate of proton permeation through the barrier. What conclusion concerning the mechanism of proton migration can be calculated from this value?

12 Figures 5.12.1 and 5.12.2 show the partial molar heat capacities of the proton in an aqueous solution, and its partial molar volume over the same temperature range.

Derive a statistical mechanical equation for the heat capacity of the proton in solution in terms of the partition functions of the various

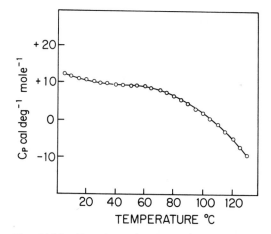

Fig. 5.12.1. Experimental values of the apparent molal heat capacity of the proton in aqueous solution as a function of temperature. (Data from Conway, 1964.)

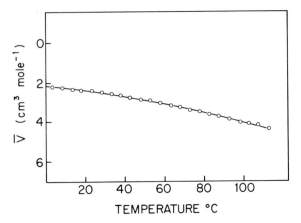

Fig. 5.12.2.   Experimental values of the apparent molal volume of the proton in aqueous solution as a function of temperature. (Data from Conway, 1964.)

bonds which H^+ forms with its surrounding water sheath. Attempt to make the model which you suggest consistent with the data given. Make a plot of your final relation between heat capacity and temperature.

Interpret the results in such a way that information on the vibrational frequencies of the solvated proton is obtained. Discuss whether the O–H bond has a continuum of vibrational energy levels in its interaction in H_3O^+ and with other waters.

ANSWERS

1   When a proton dissolves in water, there are essentially three contributions to the heat of hydration of the proton ΔH_{H^+-h}. These contributions are:

(a) The heat of the reaction:

$$H^+ + H_2O \rightarrow H_3O^+ \qquad (\Delta H_{H^+-H_2O}) \qquad (5.1.1)$$

which is the proton affinity of water.

(b) The heat of interaction of H_3O^+ with three water molecules, $\Delta H_{H_3O^+-w}$, to form the hydrated species $H_9O_4^+$. This interaction energy is given by the sum of the interactions of the ion with three water dipoles of moment μ_w, and the interaction with three water quadrupoles of moment p_w :

$$\Delta H_{H_3O^+-w} = -\frac{3N_A z_i e_0 \mu_w}{(r_{H_3O^+} + r_w)^2} + \frac{3N_A z_i e_0 p_w}{2(r_{H_3O^+} + r_w)^3} \qquad (5.1.2)$$

(c) The heat of interaction of the species $H_9O_4^+$ with the solvent, ΔH_{BC}. Regarding the solvent as a continuous dielectric medium, this term is given by the heat of Born charging a sphere of radius equal to that of $H_9O_4^+$ to a charge $+e_0$, in a medium of dielectric constant ϵ_w :

$$\Delta H_{BC} = -\frac{N_A e_0^2}{2(r_{H_3O^+} + 2r_w)} \left(1 - \frac{1}{\epsilon_w} - \frac{T}{\epsilon_w^2} \frac{\partial \epsilon_w}{\partial T}\right) \qquad (5.1.3)$$

Consequently,

$$\Delta H_{H^+-h} = \Delta H_{H^+-H_2O} + \Delta H_{H_3O^+-w} + \Delta H_{BC} \qquad (5.1.4)$$

Substituting numerical values into (5.1.2), we have

$$\Delta H_{H_3O^+-w} = -\frac{6.023 \times 10^{23} \times 3 \times 4.8 \times 10^{-10} \times 1.87 \times 10^{-18}}{(1.05 + 1.38)^2 \times 10^{-16}}$$

$$+ \frac{6.023 \times 10^{23} \times 4.8 \times 10^{-10} \times 3.9 \times 10^{-26} \times 3}{2(1.05 + 1.38)^3 \times 10^{-24}}$$

$$= -27.46 \times 10^{11} + 11.79 \times 10^{11} \text{ ergs mole}^{-1}$$

$$= -65.62 + 28.16 \text{ kcal mole}^{-1}$$

Substituting numerical values into (5.1.3), we have

$$\Delta H_{BC} = -\frac{6.023 \times 10^{23} \times (4.8)^2 \times 10^{-20}}{2(1.05 + 2.76)} \left(1 - \frac{1}{78.5} + \frac{298 \times 0.356}{(78.5)^2}\right)$$

$$= -18.29 \times 10^{11} \text{ ergs mole}^{-1}$$

$$= -43.72 \text{ kcal mole}^{-1}$$

Substituting for ΔH_{H^+-h}, $\Delta H_{H_3O^+-w}$, and ΔH_{BC} in (5.1.4) yields

$$\Delta H_{H^+-H_2O} = -263 + 65.6 - 28.2 + 43.7 \text{ kcal mole}^{-1}$$

$$= -181.9 \text{ kcal mole}^{-1}$$

| 3 | The proton affinity of water is the heat change of the reaction

$$H_2O + H^+ \rightarrow H_3O^+ \qquad (\Delta H_{H^+-H_2O}) \qquad (5.3.1)$$

when it occurs in the gas phase. Similarly, the deuteron affinity of water is the heat change of the reaction

$$H_2O + D^+ \rightarrow DH_2O^+ \qquad (\Delta H_{D^++H_2O}) \qquad (5.3.2)$$

when it occurs in the gas phase.

At $0°K$, the difference between the deuteron and proton affinities of water, $\Delta H_{(D^+-H^+)H_2O}$, is given by

$$\Delta H_{(D^+-H^+)H_2O} = \left[\sum_1^6 (\tfrac{1}{2}h\nu_0)_{DH_2O^+} - \sum_1^3 (\tfrac{1}{2}h\nu_0)_{H_2O} \right]$$

$$- \left[\sum_1^6 (\tfrac{1}{2}h\nu_0)_{H_3O^+} - \sum_1^3 (\tfrac{1}{2}h\nu_0)_{H_2O} \right] \qquad (5.3.3)$$

where the summations are the total zero-point energies of the subscripted species. Equation (5.3.3) is valid because the electronic contributions to the total energies of H_3O^+ and DH_2O^+ are constant and, at $0°K$, the rotational and translational contributions to the total energy vanish.

Equation (5.3.3) can be written

$$\Delta H_{(D^+-H^+)H_2O} = \sum_1^6 (\tfrac{1}{2}h\nu_0)_{DH_2O^+} - \sum_1^6 (\tfrac{1}{2}h\nu_0)_{H_3O^+} \qquad (5.3.4)$$

Substituting numerical values in (5.3.4) from Table 5.3.1 yields

$$\Delta H_{(D^+-H^+)H_2O} = 0.5 \times 6.623 \times 10^{-27} \times (13.76 - 15.06) \times 10^3$$

$$= -4.30 \times 10^{-24} \text{ erg sec cm}^{-1}$$

Multiplying by Avogadro's number and the velocity of light (in cm sec^{-1}), we have

$$\Delta H_{(D^+-H^+)H_2O} = -4.30 \times 10^{-24} \times 6.023 \times 10^{23} \times 3.0 \times 10^{10}$$

$$= -77.7 \times 10^9 \text{ ergs mole}^{-1}$$

$$= -1.85 \text{ kcal mole}^{-1}$$

since $\Delta H_{H^+-H_2O}$ is equal to -182 kcal mole^{-1}, then

$$\Delta H_{D^+-H_2O} = -184 \text{ kcal mole}^{-1}$$

---

5 If we assume a water molecule to be a free, classical rotator, then

$$\tfrac{1}{2}I\omega^2 = \tfrac{1}{2}kT \tag{5.5.1}$$

where I is the moment of inertia of the water molecule and ω the angular frequency of rotation. Substituting the given value for I in (5.5.1), we have

$$\omega = (1.38 \times 10^{-16} \times 298/2.0 \times 10^{-40})^{1/2}$$
$$= 1.43 \times 10^{13} \text{ rad sec}^{-1}$$

The fact that the measured dielectric relaxation time is much longer than the time of rotation of a free water molecule confirms that liquid water does not consist of free molecules. Hydrogen bonding between water molecules gives liquid water a structure similar to a somewhat broken-down and slightly expanded form of the ice structure.

Consider an aqueous solution, of molarity M, of a completely dissociated monobasic acid. The number of protons in a lamina 1 cm^2 in cross-sectional area and one water molecule thick is

$$2r_w M N_A/1000$$

where r_w is the radius of a water molecule in centimeters. Since the actual rate of proton exchange transfer (velocity) in either direction is given as 7.5×10^5 cm sec^{-1}, the time for the protons to pass through the lamina is $2r_w/(7.5 \times 10^5)$ sec. Consequently, the number of protons passing through the lamina in 1 sec is

$$2r_w M N_A \times 7.5 \times 10^5/1000 \times 2r_w$$

Since there are $2r_w N_A \times 55.5/1000$ water molecules in the lamina, then the number of protons passing through one water molecule per second is

$$M N_A \times 7.5 \times 10^5/55.5 \times 2r_w N_A$$

which on substitution for r_w becomes

$$M \times 7.5 \times 10^5/55.5 \times 2.8 \times 10^{-8} = M \times 4.87 \times 10^{11} \text{ sec}^{-1}$$

If water rotation has an energy of activation of 10 kcal mole^{-1} resulting from the energy to break hydrogen bonds, then the angular frequency of rotation of a water molecule in the structured liquid is given by

$$\omega_b = \omega_{fr}\exp(-10{,}000/RT)$$

where ω_b and ω_{fr} are the angular frequencies of the bound and freely rotating water molecules, respectively. Substituting 1.43×10^{13} rad sec^{-1} for ω_{fr}, we have

$$\begin{aligned}\omega_b &= 1.43 \times 10^{13}\exp(-10{,}000/1.987 \times 298)\\ &= 8.82 \times 10^5 \text{ rad sec}^{-1}\\ &= 1.40 \times 10^5 \text{ sec}^{-1}\end{aligned}$$

Except at low proton concentrations, about 10^{-7} or less, the rate of proton transfer, $M \times 4.87 \times 10^{11}$ sec^{-1}, is much greater rate of rotation of the water molecule. Consequently, it is the latter step that must be rate-determining for proton migration.

$\boxed{7}$ In pure water or ice, the hydrogen ion concentration is given by

$$C_{\text{H}^+} = 10^{-pK_w/2} \tag{5.7.1}$$

Substituting the given values of the pK_w for ice and water at $-10°C$ into (5.7.1), we have

$$C_{\text{H}^+-\text{ice}} = 5.0 \times 10^{-11} \text{ mole liter}^{-1}$$

and

$$C_{\text{H}^+-\text{water}} = 2.8 \times 10^{-8} \text{ mole liter}^{-1}$$

The hydroxide ion concentrations are, of course, equal to the hydrogen ion concentrations.

The specific conductivity J of a 1:1 electrolyte solution is given by

$$J = CF(u_+ + u_-) \tag{5.7.2}$$

where u_+ and u_- are the conventional mobilities of the cation and anion, respectively, and C is the concentration of the electrolyte in moles cm^{-3}. Substituting for J and C in (5.7.2), we have

$$(u_+ + u_-)_{ice} = 1.4 \times 10^{-9}/5.0 \times 10^{-14} \times 96500$$
$$= 0.290 \text{ cm}^2 \text{ sec}^{-1} \text{ V}^{-1}$$

and

$$(u_+ + u_-)_{water} = 1.4 \times 10^{-9}/2.8 \times 10^{-11} \times 96500$$
$$= 5.2 \times 10^{-4} \text{ cm}^2 \text{ sec}^{-1} \text{ V}^{-1}$$

Since, for a 1:1 electrolyte, the transport number t_+ is given by

$$t_+ = u_+/(u_+ + u_-) \tag{5.7.3}$$

we have, by making use of the given value of t_+,

$$u_{+-ice} = 0.290 \times 0.64$$
$$= 0.185 \text{ cm}^2 \text{ sec}^{-1} \text{ V}^{-1}$$

and

$$u_{+-water} = 5.2 \times 10^{-4} \times 0.64$$
$$= 3.3 \times 10^{-4} \text{ cm}^2 \text{ sec}^{-1} \text{ V}^{-1}$$

Consequently, although the specific conductivities of ice and water are almost the same, the mobility of the proton in ice is about 10^3 times that in water.

In Problem 5, it was shown that the rate of proton transfer per water molecule is $M \times 4.87 \times 10^{11} \text{ sec}^{-1}$, where M is the molarity of the protons. Hence, as the concentration of protons is lowered, any given water molecule then has passing across it, along a given coordinate, a decreasing number of protons per second. The number of times per second that a typical water molecule undergoes field-induced rotation, so that it can receive the tunneling proton, is reduced as the concentration of protons is reduced. At low concentration of protons, the rate of rotation of the water molecule needed to pass on the proton drops to less than the rate of thermal rotation of the molecule. Proton tunneling is then the rate-determining step.

The rate of proton transfer per water molecule in ice is $4.87 \times 10^{11} \times 5 \times 10^{-11} \text{ sec}^{-1}$. Since this value is much smaller than the rate of spontaneous thermal rotation in ice, about 10^6 sec^{-1}, proton tunneling is rate-determining in ice.

9 Looking at the left-hand scale of Fig. 5.9.1, we have that the pfeI of acetic acid and chloroacetic acid are -0.385 and -0.275 eV, respectively. The left-hand scale of Fig. 5.9.1 actually gives the standard free energy of an occupied proton level in the acid with respect to the standard free energy of the occupied proton level in H_3O^+. Consequently, for the process

$$HA + H_2O \rightleftarrows H_3O^+ + A^- \tag{5.9.1}$$

where HA is an acid and A^- is its conjugate base,

$$-U_{HA} = \Delta G^\circ_{HA} \tag{5.9.2}$$

where U_{HA} is the proton free energy level of HA and ΔG°_{HA} is the standard free energy change of (5.9.1). Using the conversion factor

$$1 \text{ eV molecule}^{-1} = 23.06 \text{ kcal mole}^{-1}$$

we have that

$$\Delta G^\circ_{AcOH} = 8.87 \text{ kcal mole}^{-1}$$

$$\Delta G^\circ_{Cl-AcOH} = 6.63 \text{ kcal mole}^{-1}$$

The acid dissociation (5.9.1) is traditionally written in the form

$$HA \rightleftarrows A^- + H^+ \tag{5.9.3}$$

and the dissociation constant K_A is defined as

$$K_A = a_{A^-}a_{H^+}/a_{HA} \tag{5.9.4}$$

The equilibrium constant K of (5.9.1) is given by

$$K = a_{A^-}a_{H_3O^+}/a_{HA}a_{H_2O} \tag{5.9.5}$$

Neglecting activity coefficients, (5.9.4) and (5.9.5) may be written in terms of the corresponding concentration terms. Consequently, K_A and K will be numerically equal only for dilute solution and when the concentrations are expressed as mole fractions. In terms of the usual molar units of concentration, we have

$$K_A = KC_{H_2O} \tag{5.9.6}$$

since the protons referred to in equation (5.9.3) are bound to water molecules and therefore

$$C_{H^+} = C_{H_3O^+} \qquad (5.9.7)$$

Since

$$K = \exp(-\Delta G^\circ / RT) \qquad (5.9.8)$$

$$K_A = C_{H_2O} \exp(-\Delta G^\circ / RT) \qquad (5.9.9)$$

Substituting for ΔG° and taking C_{H_2O} as 55.5 moles liter^{-1}, we have

$$(K_A)_{\text{acetic}} = 1.75 \times 10^{-5} \, M$$
$$(K_A)_{\text{chloroacetic}} = 1.26 \times 10^{-3} \, M$$

and since

$$pK_A = -\log_{10} K_A$$
$$(pK_A)_{\text{acetic}} = 4.76$$
$$(pK_A)_{\text{chloroacetic}} = 2.90$$

The relative strength of one acid HA_1 with respect to that of another HA_2 in a given solvent is defined by the equilibrium constant K' for the reaction

$$HA_1 + A_2^- \rightleftharpoons HA_2 + A_1^- \qquad (5.9.10)$$

The larger the value of K', the greater the relative strength of HA_1 with respect to HA_2. Now,

$$K' = a_{HA_2} a_{A_1^-} / a_{HA_1} a_{A_2^-} \qquad (5.9.11)$$

$$= K_{A_1} / K_{A_2} \qquad (5.9.12)$$

Substituting the given values for chloroacetic and β-chloropropionic acids into (5.9.12) yields

$$K' = 1.3 \times 10^{-3} / 1.0 \times 10^{-4}$$
$$= 13$$

The equilibrium constant K' can be written in terms of the standard free energy change of (5.9.10), $\Delta G^{\circ\prime}$, as

$$-RT \ln K' = \Delta G^{\circ\prime} \qquad (5.9.13)$$

Now, $\Delta G^{\circ\prime}$ can be separated into two terms, a chemical term ΔG_C° and an electrostatic term. Since A_1^- and A_2^- are the only charged species participating in (5.9.10), the electrostatic term is simply the difference between the charging energies of A_1^- and A_2^-. Hence,

$$\ln K' = -\frac{\Delta G_C^{\circ}}{RT} - \frac{N_A e_0^2}{2\epsilon RT}\left(\frac{1}{r_{A_1^-}} - \frac{1}{r_{A_2^-}}\right) \qquad (5.9.14)$$

If K' is known in one solvent, $\Delta G_C^{\circ}/RT$ can be evaluated and then K' calculated in another solvent. Evaluating $\Delta G_C^{\circ}/RT$ for the given data in aqueous solution, we have

$$-\frac{\Delta G_C^{\circ}}{RT} = \ln K' + \frac{N_A e_0^2}{2\epsilon RT}\left(\frac{1}{r_{A_1^-}} - \frac{1}{r_{A_2^-}}\right) \qquad (5.9.15)$$

and substituting numerical values into (5.9.15) yields

$$-\frac{\Delta G_C^{\circ}}{RT} = \ln(13) + \frac{6.023 \times 10^{23} \times (4.8)^2 \times 10^{-20} \times 10^{-7}}{2 \times 78.5 \times 8.314 \times 298 \times 10^{-8}}$$

$$\times \left(\frac{1}{2.9} - \frac{1}{4.2}\right)$$

$$= 2.56 + 0.38$$

$$= 2.94$$

Substituting this value in (5.9.14) and taking the dielectric constant of methanol as 32.5 yields

$$\ln K' = 2.94 - \frac{6.023 \times 10^{23} \times (4.8)^2 \times 10^{-20} \times 10^{-7}}{2 \times 32.5 \times 8.314 \times 298 \times 10^{-8}}\left(\frac{1}{2.9} - \frac{1}{4.2}\right)$$

$$= 2.94 - 0.92$$

$$\bar{K}' = 7.5$$

Therefore, the relative strength of chloroacetic acid to that of β-chloropropionic acid in methanol is 7.5. It should be noted that the above procedure is valid only for similar acids and similar solvents; only then can the chemical term ΔG_C° be regarded as constant.

CHAPTER 6

IONIC LIQUIDS

1 The probability that a hole in a liquid will have a radius between r and $r + dr$ is $P\, dr$ and is given by

$$P\, dr = (16a^{7/2}r^6/15\pi^{1/2}) \exp(-ar^2)\, dr \qquad (6.1.1)$$

where

$$a = 4\pi\gamma/kT \qquad (6.1.2)$$

and γ is the surface tension of the liquid. Show that the mean hole volume v_h is given by

$$v_h = 0.68(kT/\gamma)^{3/2} \qquad (6.1.3)$$

Use the data of Table 6.1.1 to calculate the mean size of holes in molten salts and compare these values with the volumes of the ions concerned. Comment on the conclusion.

TABLE 6.1.1. Surface Tension and Ionic Radii of Various Molten Salts

Salt	T, °C	γ, dynes cm^{-1}	r_+, Å	r_-, Å
NaCl	820	115	0.95	1.81
NaBr	785	97	0.95	1.95
NaI	670	87	0.95	2.16
KCl	800	97	1.33	1.81
KBr	750	88	1.33	1.95
KI	720	76	1.33	2.16
CaCl$_2$	585	83	1.21	1.81

2 Table 6.2.1 lists the molar volume change on fusion of various salts together with the free volume per mole calculated from ultrasonic compressibility measurements. What model of molten salts do the data support?

The free volume and hole volume of some molten salts obtained at various temperatures are listed in Table 6.2.2. Plot the free volume against the hole volume and make a physical interpretation of the result.

TABLE 6.2.1. Molar Volume Change of Fusion, ΔV, and the Molar Free Volume, V_f, of Various Salts

Salt	ΔV, ml mole^{-1}	V_f, ml mole^{-1}
LiCl	5.9	1.0
NaCl	7.4	0.75
NaBr	8.2	1.15
NaI	8.7	1.15
KCl	7.1	0.85
KBr	8.0	1.02
KI	9.4	0.95
CsCl	13.3	0.78
CsBr	14.3	0.86

TABLE 6.2.2. Molar Free Volume V_f and Molar Hole Volume V_h for Molten Salts

Salt	V_f, ml mole^{-1}	V_h, ml mole^{-1}
LiCl	0.36	10.8
	0.63	17.5
	0.78	21.7
	0.96	25.9
NaBr	0.51	20.5
	0.57	25.3
	0.75	31.3
	0.93	38.5
KI	0.45	28.3
	0.63	40.3
	0.90	56.6
	1.29	71.7

[3] Assuming the Stokes–Einstein relation, calculate the anionic transport numbers from the data listed in Table 6.3.1. Discuss the results.

Derive a relation, making use of the Stokes–Einstein relation, for the anionic transport number of a fused 2:1 electrolyte taking into account the dissociation equilibria

$$MX_2 \rightleftarrows MX^+ + X^- \qquad (6.3.1a)$$

$$MX^+ \rightleftarrows M^{2+} + X^- \qquad (6.3.1b)$$

Using the Pauling radii ($r_{PbCl^+} = 4.83$ Å; $r_{Pb^{2+}} = 1.21$ Å; $r_{Cl^-} = 1.81$ Å) and the experimental anionic transport number of 0.8, determine whether Pb^{2+} or $PbCl^+$ is the predominant cationic species in fused lead chloride.

TABLE 6.3.1. Experimental Anionic Transport Numbers and Ionic Radii

Salt	r_+, Å	r_-, Å	t_-
LiCl	0.60	1.81	0.25
NaCl	0.95	1.81	0.38
KCl	1.33	1.81	0.38
RbCl	1.48	1.81	0.42

[4] In the Temkin model of an ideal solution of molten salts, each salt is considered to be completely ionized. Further, if the physical properties of all the cations and all the anions, respectively, are considered to be exactly the same, the total enthalpy and energy of the solution will be that of the pure components. Show that the entropy of mixing ΔS_{mix} is given by

$$-\Delta S_{mix}/R = \sum n_i^+ \ln N_i^+ + \sum n_i^- \ln N_i^- \qquad (6.4.1)$$

where n_i^+ and n_i^- are the numbers of gram ions of cations and anions, respectively, from component i of the solution, and N_i^+ and N_i^- are the ion fractions defined by

$$N_i^+ = n_i^+ / \sum n_i^+, \qquad N_i^- = n_i^- / \sum n_i^-$$

In fused $NaNO_3$–KNO_3 at 263°C, Cd^{++} and Br^- form a series
of complexes. Transition time measurements can, in this case,
be used to obtain information on the transformation of these complexes. The transition time measurements are made by applying
cathodic galvanostatic pulses to a hanging mercury drop electrode
in the melt. The overall process is known to be

$$CdBr_3^- \rightleftarrows CdBr_2 + Br^- \qquad (6.5.1a)$$

$$CdBr_2 \rightleftarrows CdBr^+ + Br^- \qquad (6.5.1b)$$

$$CdBr^+ \rightleftarrows Cd^{++} + Br^- \qquad (6.5.1c)$$

$$Cd^{++} + 2e^- \rightleftarrows Cd(Hg) \qquad (6.5.1d)$$

The formation constants K_1, K_2, and K_3 corresponding to steps
(a), (b), and (c) are known to be 8 ± 3, 65 ± 33, and 100 ± 50
kg mole^{-1}, respectively.

For processes represented by the reaction scheme (6.5.1), the
transition time τ is given by

$$\tau^{1/2} = \frac{\pi^{1/2}D^{1/2}nFC}{2i} - \frac{\pi^{1/2}KC_{Br^-}}{2(k_+ + k_-C_{Br^-})^{1/2}} \, \mathrm{erf}(Z) \qquad (6.5.2)$$

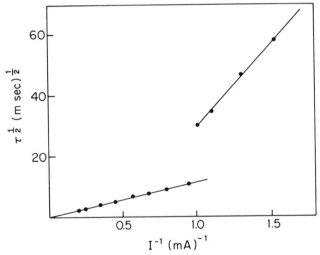

Fig. 6.5.1. Results of transition time experiments, $\tau^{1/2}$ plotted
against I^{-1} for a $NaNO_3$–KNO_3 melt containing 4.82×10^{-3} mole
kg^{-1} Cd^{2+} and 9.21×10^{-2} mole kg^{-1} KBr.

where $Z = (k_+ + k_- C_{Br^-})^{1/2} \tau^{1/2}$, with C_{Br^-} the bulk concentration of Br^- and C the sum of the bulk concentrations of all species containing Cd^{++}; i is the cathodic current density, and K, k_+, and k_- refer respectively, to the formation constant, the forward rate constant, and the reverse rate constant of the rate-determining step. Devise a means of calculating K, k_+, and k_-. To which step in the overall process do these quantities refer ?

Results of experiments are shown in Fig. 6.5.1, for which $C_{Br^-} = 9.21 \times 10^{-2}$ mole kg^{-1} and $C = 4.82 \times 10^{-3}$ mole kg^{-1}.

| 6 | (a) Transition times for silver deposition from fused $NaNO_3$ at 330°C are listed for various melt compositions in Table 6.6.1; a cathodic pulse of constant current density is being used. Calculate the solubility product of silver chloride under these conditions.

(b) The data of Table 6.6.2 refer to a constant melt composition of 2.01×10^{-2} M $AgNO_3$ in fused $NaNO_3$ at various temperatures. Calculate the activation energy and preexponential factor for diffusion of silver ions in fused $NaNO_3$ over the temperature range 280–317°C.

TABLE 6.6.1.   Transition Times for Silver Deposition from $NaNO_3$ Melts of Various Compositions at 330°C

$10^2 C_{AgNO_3}$, molal	$10^2 C_{NaCl}$, molal	τ, msec
2.49	0	805
	0.52	805
	0.93	650
	1.83	400
	2.44	275
	2.93	210
	3.77	155

| 7 | In the "density fluctuation hole model" of molten salts, the holes are thermally distributed in size. Hole formation in these liquids can be formally treated as a chemical reaction and

$$N_h/N = \exp(-\Delta G_h^\circ / RT) \qquad (6.7.1)$$

**TABLE 6.6.2.   Transition Times for Silver Deposition
from NaNO$_3$ Melts of Composition 2.01 × 10^{-2} M AgNO$_3$
at Various Temperatures and Current Densities**

Temp., °C	Current density, mA cm^{-2}	Transition time, msec
317	10	470
	7	960
	5	1880
	3	5220
297	10	383
	7	780
	5	1530
	3	4250
280	10	324
	7	660
	5	1290
	3	3600

where N_h and N are the equilibrium numbers of holes and ions per mole of liquid, respectively, and $\Delta G_h{}^\circ$ is the standard free energy of hole formation. The diffusion process can be visualized as the net result of a density fluctuation in the salt which momentarily provides a local cavity and the jumping of a sufficiently activated ion into the cavity. If the probability of a successful collision between an ion and an adjacent hole is given by $\exp(-\Delta G_j{}^*/RT)$, where $\Delta G_j{}^*$ is the free energy for jumping, show that

$$E_p = \Delta H_h{}^\circ + \Delta E_j{}^* \tag{6.7.2}$$

where E_p is the activation energy for diffusion at constant pressure, $\Delta H_h{}^\circ$ is the standard enthalpy of hole formation, and $\Delta E_j{}^*$ is the activation energy for the jump of an ion into a hole.

Utilizing the information that the change in volume of a melt with temperature is predominantly due to the change in the number of holes and only in a small part due to the change in the most probable hole volume, rationalize the relation

$$E_v = \Delta E_j{}^* \tag{6.7.3}$$

where E_v is the energy of activation for diffusion at constant volume.

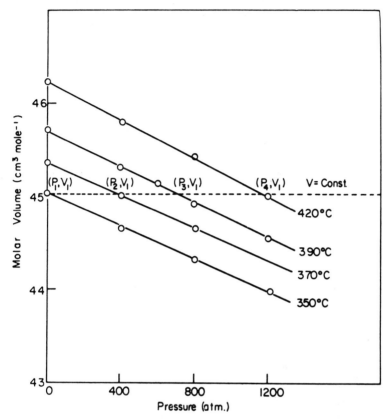

Fig. 6.7.1. Plot of the molar volume of molten $NaNO_3$ as a function of pressure at various temperatures. (Data from Nagarajan and Bockris, 1966.)

Using the data of Fig. 6.7.1 and Table 6.7.1, calculate E_p and E_v for sodium diffusion in fused sodium nitrate. Discuss whether or not the values you calculate are consistent with the density fluctuation hole model.

8 Electrical conductance, molar volume, surface tension, UV spectra, vapor pressure, and self-diffusion studies on molten $CdCl_2$–KCl mixtures all indicate the formation of complex ions. Tetrahedral $CdCl_4^{2-}$, planar triangular $CdCl_3^-$, and pyramidal $CdCl_3^-$

TABLE 6.7.1.   Effect of Pressure and Temperature on the Diffusion Coefficient of ^{22}Na in NaNO$_3$*

Pressure, atm	10^5 Diffusion coefficient, cm^2 sec^{-1}			
	350°C	370°C	390°C	420°C
1	2.14	2.38	2.61	3.04
400	1.91	2.16	2.32	2.65
600	—	—	2.21	—
800	1.74	1.99	2.08	2.50
1200	1.58	—	—	—

* Data from Nagarajan and Bockris, 1966.

have been suggested as the most probable complex ions in this melt. The complex ion formation has been studied by Raman spectroscopy and Table 6.8.1 summarizes the Raman spectra attributed to the complex ion.

What are the point groups for the complex ions suggested above and how many Raman frequencies would you expect for each complex ion and what is the state of polarization of each? Compare the results of these considerations with the data of Table 6.8.1 and decide which is the predominant complex ion formed under the conditions of the Raman study.

TABLE 6.8.1.   Summary of Raman Spectroscopic Data for Molten CdCl$_2$–KCl Mixture

Composition, mole % KCl	Temperature range, °C	Frequency $\Delta\nu$, cm^{-1}	Intensity	State of polarization
50	482–708	257 ± 2	Strong	Polarized
		245 ± 3	Weak	Depolarized
		211 ± 6	Weak	Polarized
		177 ± 5	Very weak	Depolarized

9 | Show that for a diffusion process in a molten salt,

$$\Delta V^* = -RT(\partial \ln D/\partial P)_T \tag{6.9.1}$$

where D is the diffusion coefficient and ΔV^* the activation volume. Further, show that for the density fluctuation hole model,

$$\Delta V^* = V_h + \Delta V_j^* \qquad (6.9.2)$$

where V_h is the most probable molar volume of holes in the melt and ΔV_j^* is the activation volume for the jumping process.

Using the data of Tables 6.7.1 and 6.9.1 and Fig. 6.7.1, calculate the activation volume for diffusion of Na^{22} and Cs^{134} in fused sodium nitrate. Calculate ΔV_j^* for each diffusion process and comment on the numerical values of these terms. Are they consistent with the density fluctuation hole model? Take the surface tension of fused $NaNO_3$ as 110 dynes cm^{-1} at 390°C.

TABLE 6.9.1. **Effect of Pressure and Temperature on the Diffusion Coefficient of ^{134}Cs in Fused $NaNO_3$***

Pressure, atm	10^5 Diffusion coefficient, cm² sec⁻¹			
	350°C	370°C	390°C	420°C
1	2.31	2.62	2.90	3.39
400	2.05	2.35	2.64	3.03
800	1.83	2.07	2.42	—

* Data from Nagarajan and Bockris, 1966.

---

10 The following facts are known about simple liquid silicates:

(a) The heat of activation for viscous flow is very high (134 kcal mol⁻¹) for pure SiO_2 (Fig. 6.10.1). However, during the addition of the first 10 mole % M_2O, the heat of activation for viscous flow drops to about 40 kcal mole⁻¹ and thereafter decreased only slightly on the addition of further M_2O down to ∼50 mole % M_2O.

(b) Compressibility and expansivity of simple liquid silicates are negligible from 0 to 10 % M_2O. At this composition, they undergo a sudden inflection and rise sharply with increasing M_2O content (Fig. 6.10.2).

(c) The partial molar volume of SiO_2 in all compositions of M_2O–SiO_2 is independent of composition.

Discuss in depth whether these facts are more consistent with the

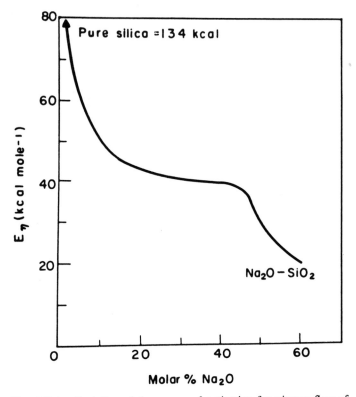

Fig. 6.10.1. Variation of the energy of activation for viscous flow of a $Na_2O + SiO_2$ melt as a function of mole percentage of Na_2O.

classical three-dimensional theory of glasses, due to Zachariasen (1932), or with a model in which, at $>10\% M_2O$, discrete silicate anions become the predominant structural entity of silicates and glasses.

|11| Figure 6.11.1 shows a plot of the heat of activation at constant pressure for viscous flow, E_η, of a number of ionic liquids plotted against melting point T_m. Rationalize the relation

$$E_\eta = 3.7RT_m \qquad (6.11.1)$$

according to some model of fused salts.

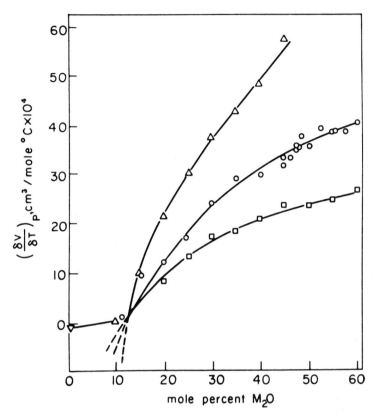

Fig. 6.10.2. The sharp change in the expansivity of M_2O-SiO_2 melts around 10 mole % M_2O composition; ($\triangle$) $K_2O + SiO_2$; ($\circ$) $Na_2O + SiO_2$; ($\square$) $Li_2O + SiO_2$; ($\triangledown$) SiO_2 .

| 12 | A microdensitometer trace of the Raman spectrum of molten cryolite at 1030°C is shown in Fig. 6.12.1(a), and in Fig. 6.12.1(b), it is analyzed into its thermal radiation, instrumental background, and Raman scattering components. Only two Raman frequencies, one centered at 633 and the other at 575 Δcm^{-1}, are observed, both being polarized. The band at 575 Δcm^{-1} is very weak.

The spectrum of solid cryolite at room temperature, which is independent of crystal orientation, is shown in Fig. 6.12.2. X-ray diffraction has shown that crystalline cryolite contains finite AlF_6^{3-}

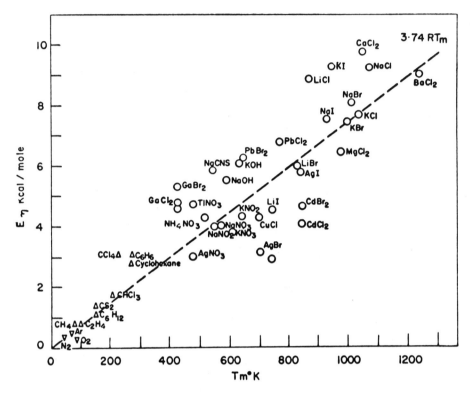

Fig. 6.11.1.   The dependence of the experimental energy of activation for viscous flow on the melting point. (Data from Nanis and Bockris, 1963.)

octahedra. However, there is probably some distortion of the ideal cubic structure at room temperature.

From symmetry considerations and comparison with the Raman spectra of tetrahedral and octahedral fluorides, show that it is reasonable to assign the bands at 633 and 575 Δcm^{-1} to AlF_4^- and AlF_6^{3-}, respectively. On the basis that the principal dissociation in the melt is

$$AlF_6^{3-} \rightleftarrows AlF_4^- + 2F^- \qquad (6.12.1)$$

roughly estimate the dissociation constant.

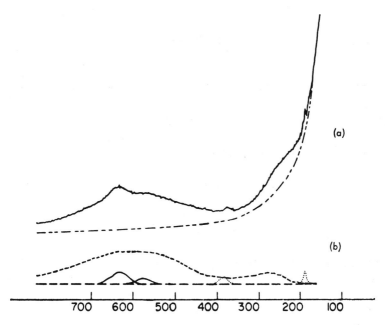

Fig. 6.12.1. (a) Spectrum of molten cryolite at 1030°C. (– - - –) Halation. (b) Analysis of the spectrum into Raman scattering (——), thermal radiation (– – –), and instrumental background radiation (- - -). (Data from Solomon *et al.*, 1968.)

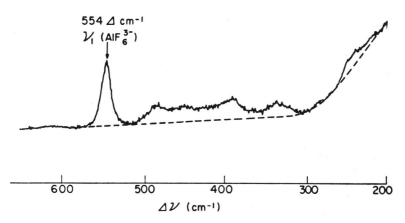

Fig. 6.12.2.   Raman spectrum of single-crystal cryolite at 30°C. (Data from Solomon *et al.*, 1968.)

ANSWERS

1   Assuming spherical holes, the mean hole volume is given by

$$v_h = 4\pi\langle r^3\rangle/3 \tag{6.1.4}$$

where, from (6.1.1), $\langle r^3\rangle$ is given by

$$\langle r^3\rangle = \int_0^\infty r^3 P\, dr$$

$$= \int_0^\infty r^3 \frac{16a^{7/2}r^6}{15\pi^{1/2}} \exp(-ar^2)\, dr \tag{6.1.5}$$

and making the substitutions $t = ar^2$ and $K = 16a^{7/2}/15\pi^{1/2}$ in (6.1.5), we have

$$\langle r^3\rangle = (K/2a^5) \int_0^\infty t^4 e^{-t}\, dt \tag{6.1.6}$$

Since the integral in (6.1.6) is a gamma function of value 4!, (6.1.6) becomes

$$\langle r^3\rangle = (K/2a^5)(4!) \tag{6.1.7}$$

Substituting for K and a [Eq. (6.1.2)] in (6.1.7) yields

$$\langle r^3\rangle = (8/5\pi^2)(kT/\gamma)^{3/2} \tag{6.1.8}$$

and substituting (6.1.8) into (6.1.4) yields (6.1.3),

$$v_h = (32/15\pi)(kT/\gamma)^{3/2}$$
$$= 0.68(kT/\gamma)^{3/2}$$

For NaCl, the mean hole volume is, by substitution of numerical data into (6.1.3),

$$v_{h(\text{NaCl})} = 0.68(1.38 \times 10^{-16} \times 1093/115)^{3/2}$$
$$= 32.3 \times 10^{-24} \text{ cm}^3$$
$$= 32.3 \text{ Å}^3$$

Table 6.1.2 lists the mean hole volume of the salts together with the volumes of the individual ions. It is apparent that the mean hole volume of a molten salt is much larger than the volume of the cation and in general slightly larger than the volume of the anion. In the hole theory of fused salts, migration or diffusion of ions takes place by an ion hopping into a hole. Consequently, it is necessary for the holes to be large enough to accommodate an anion or a cation.

TABLE 6.1.2.   Mean Hole Volumes v_h Compared to the Volumes of Anions and Cations in Certain Molten Salts

Salt	v_h , Å³	v_+ , Å³	v_- , Å³
NaCl	32.3	3.59	24.8
NaBr	39.7	3.59	31.1
NaI	39.3	3.59	42.2
KCl	40.7	9.85	24.8
KBr	43.5	9.85	31.1
KI	52.1	9.85	42.2
$CaCl_2$	34.6	3.82	26.8

3

The Stokes–Einstein equation relates the mobility u_i of an ion i to the viscosity of the medium η by

$$u_i = z_i e_0 / 6\pi r_i \eta \qquad (6.3.2)$$

where r_i is the ionic radius. The transport number of the ith species is defined by

$$t_i = I_i / \sum I_j \qquad (6.3.3)$$

where the current I_j due to a particular species is given by

$$I_j = z_j F C_j u_j X \qquad (6.3.4)$$

where C_j is the concentration and X is the electric field. For a 1:1 electrolyte, the transport number of the anion t_- is, by substitution of (6.3.4) into (6.3.3),

$$t_- = u_- / (u_+ + u_-) \qquad (6.3.5)$$

and substituting (6.3.2) into (6.3.5) yields

$$t_- = r_+ / (r_+ + r_-) \qquad (6.3.6)$$

TABLE 6.3.2.   Experimental and Calculated Anionic Transport Numbers for Some Fused Salts

Salt	t_-(calc)	t_-(exp)
LiCl	0.25	0.25
NaCl	0.34	0.38
KCl	0.42	0.38
RbCl	0.45	0.42

For fused lithium chloride, the transport number of Cl^- is

$$t_- = 0.60/(0.60 + 1.81) = 0.25$$

Table 6.3.2 compares experimental anionic transport numbers with those calculated on the basis of the Stokes–Einstein relation. The agreement is seen to be good, particularly when the experimental difficulties of measuring transport numbers of fused salts is recalled and when it is further recalled that the Stokes relation, from which (6.3.2) is derived, is valid only when the moving particle is large compared to the sizes of particles that compose the medium. This is obviously not the case for ionic migration in a fused salt. However, this agreement shows that this approach can be used to obtain an approximate value of the transport number in unknown systems.

For a 2:1 electrolyte undergoing the dissociation (6.3.1), we have from Eqs. (6.3.3) and (6.3.4)

$$t_- = C_- u_- / (C_+ u_+ + C_- u_- + 2C_{2+} u_{2+}) \tag{6.3.7}$$

where the subscripts $-$, $+$, and $2+$ refers to the species X^-, MX^+, and M^{2+}, respectively. Substituting (6.3.2) into (6.3.7) yields, with rearrangement,

$$t_- = C_- r_+ r_{2+} / (C_+ r_- r_{2+} + C_- r_+ r_2 + 4C_{2+} r_- r_+) \tag{6.3.8}$$

Lead chloride must be at least partly dissociated to be conducting and we may consider three distinct cases.

Case 1: $C_+ \gg C_{2+}$ and therefore $C_+ \approx C_-$ and (6.3.8) becomes

$$t_- \approx r_+ r_{2+} / (r_- r_{2+} + r_+ r_{2+}) \tag{6.3.9}$$

and, substituting numerical values, we get

$$t_- = 0.73$$

Case 2: $C_+ \approx C_{2+}$ and therefore $C_- \approx 3C_+$ and (6.3.8) becomes

$$t_- = 3r_+r_{2+}/(r_-r_{2+} + 4r_-r_+ + 3r_+r_{2+}) \qquad (6.3.10)$$

and, substituting numerical values, we get

$$t_- \approx 0.32$$

Case 3: $C_{2+} \gg C_+$ and therefore $C_- \approx 2C_{2+}$ and (6.3.8) becomes

$$t_- = 2r_+r_{2+}/(4r_+r_- + 2r_+r_{2+}) \qquad (6.3.4)$$

and substituting numerical values, we get

$$t_- = 0.25$$

Since the experimental value of t_{Cl^-} is 0.8, it may be concluded that case 1 best represents the available data and that $PbCl^+$ is the predominant cationic species in the melt.

---

5   Provided the term $(k_+ + k_-C_{Br^-})^{1/2}$ in (6.5.3) has a suitable value, the error function in (6.5.2) can be approximated, when τ is large, as

$$\text{erf}(Z) = 1 \qquad (6.5.4)$$

and when τ is small, as

$$\text{erf}(Z) = 2Z/\pi^{1/2} \qquad (6.5.5)$$

and Eq. (6.5.2) becomes under these limiting conditions

$$\tau^{1/2} = \frac{\pi^{1/2}D^{1/2}nFC}{2i} - \frac{\pi^{1/2}KC_{Br^-}}{2(k_+ + k_-C_{Br^-})^{1/2}} \qquad \tau \text{ large} \quad (6.5.6)$$

$$\tau^{1/2} = \frac{\pi^{1/2}D^{1/2}nFC}{2(1 + KC_{Br^-})i} \qquad \tau \text{ small} \quad (6.5.7)$$

Consequently, a plot of $\tau^{1/2}$ against i^{-1}, for a wide range of current densities, can be expected to exhibit two linear portions corresponding to the limiting conditions (6.5.6) and (6.5.7). The slopes of these two linear regions are

$$[d\tau^{1/2}/d(i^{-1})]_{\tau \text{ large}} = \tfrac{1}{2}\pi^{1/2}D^{1/2}nFC \qquad (6.5.8)$$

and

$$[d\tau^{1/2}/d(i^{-1})]_{\tau \text{ small}} = \pi^{1/2}D^{1/2}nFC/2(1 + KC_{Br^-}) \qquad (6.5.9)$$

and the ratio of the slopes is given by

$$[d\tau^{1/2}/d(i^{-1})]_{\tau \text{ large}}/[d\tau^{1/2}/d(i^{-1})]_{\tau \text{ small}} = (1 + KC_{Br^-}) \quad (6.5.10)$$

From the slopes of the linear portions of Fig. 6.5.1,

$$(1 + KC_{Br^-}) = 17.7/3.8$$

and

$$K = [(17.7/3.8) - 1]/0.0921$$
$$= 39.7 \text{ kg mole}^{-1}$$

From the intercept of the plot of $\tau^{1/2}$ (τ large) against i^{-1}, we have

$$\pi^{1/2}KC_{Br^-}/2(k_+ + k_-C_{Br^-})^{1/2} = 25.5 \text{ (m sec)}^{1/2}$$

Therefore,

$$\frac{\pi K^2 C_{Br^-}^2}{4(k_+ + k_-C_{Br^-})} = 0.650 \text{ sec}$$

and substituting for K and C_{Br^-} yields

$$k_+ + 0.0921k_- = 16.1 \text{ sec}^{-1}$$

Since K, the formation constant, is the reciprocal of the dissociation constant,

$$K = k_-/k_+ = 39.7 \text{ kg mole}^{-1}$$

Consequently,

$$k_+ = 3.5 \text{ sec}^{-1}$$
$$k_- = 137 \text{ kg mole}^{-1} \text{ sec}^{-1}$$

Since the value of K is within the range of experimental values for step (b) of (6.5.1), this step is the rate-determining step.

<table>
<tr><td>7</td></tr>
</table>

The diffusion coefficient is proportional to the product of the number of holes times the probability that an ion will jump into a hole, $\exp(-\Delta G_j^*/RT)$. From Eq. (6.7.1) and the probability of jumping, we have

$$D = D_0' \exp(-\Delta G_h^\circ/RT) \exp(-\Delta G_j^*/RT) \tag{6.7.4}$$

$$= D_0' \exp[(-\Delta H_h^\circ + T \Delta S_h^\circ - \Delta H_j^* + T \Delta S_j^*)/RT] \tag{6.7.5}$$

where the subscripts h and j refer to hole formation and jumping, respectively. Since there is no PV work performed in the jumping of an ion into an already created hole,

$$\Delta H_j^* = \Delta E_j^* \tag{6.7.6}$$

Substituting (6.7.6) into (6.7.5) yields

$$D = D_0' \exp\left(\frac{\Delta S_h^\circ + \Delta S_j^*}{R}\right) \exp\left(-\frac{\Delta H_h^\circ + \Delta E_j^*}{RT}\right) \tag{6.7.7}$$

and comparing (6.7.7) with the empirical relation

$$D = D_0 \exp(-E_p/RT) \tag{6.7.8}$$

where $D_0 \neq f(T)$ and E_p is the experimental energy of activation at constant pressure, reveals that

$$E_p = \Delta H_h^\circ + \Delta E_j^* \tag{6.7.2}$$

Since the mean hole volume v_h is approximately temperature-independent, we can write

$$V_{T_2} - V_{T_1} = v_h(N_{h(T_2)} - N_{h(T_1)}) \tag{6.7.9}$$

and substituting Eq. (6.7.1), we have

$$V_{T_2} - V_{T_1} = Nv_h[\exp(-\Delta G_h^\circ/RT_2) - \exp(-\Delta G_h^\circ/RT_1)] \tag{6.7.10}$$

but (6.7.10) can only be true at constant volume (i.e., $V_{T_2} - V_{T_1} = 0$) if, at constant volume, $\Delta H_h^\circ = 0$. Consequently, we find (6.7.3),

$$E_v \simeq \Delta E_j^*$$

Calculation of E_p: From Eq. (6.7.8), it is seen that E_p can be obtained from the slope of the plot of $\log D$ against T^{-1} at constant

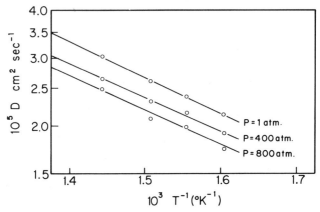

Fig. 6.7.2.   Plots of D against T^{-1} at various constant pressures for Na^{22} diffusion in fused $NaNO_3$.

pressure. Figure 6.7.2 shows such plots for 1, 400, and 800 atm. From the slope of the plot,

$$-2.303[d \log D/d(T^{-1})] = 2.17 \times 10^3 \text{ deg}$$

and

$$E_p = 1.987 \times 2170 \text{ cal mole}^{-1}$$
$$= 4.31 \text{ kcal mole}^{-1}$$

Calculation of E_v: As shown in Fig. 6.7.1, the pressure, at different temperatures, at which the melt has a constant volume V_1 is found by drawing a horizontal line through the molar volume–pressure plots. Plotting the data of Table 6.7.1 (Fig. 6.7.3), the diffusion coefficient corresponding to V_1 at different temperatures can be interpolated. E_v is then found from the slope of the plot of $\log[D_{(V_1)}]$ against (T^{-1}) as shown in Fig. 6.7.4:

$$-2.303\{d \log[D_{(V_1)}]/d(T^{-1})\} = 393 \text{ deg}$$
$$E_v = 1.987 \times 393 \text{ cal mole}^{-1}$$
$$= 0.78 \text{ kcal mole}^{-1}$$

From Eqs. (6.7.2) and (6.7.3), we have

$$\Delta H_h{}^\circ = 4.31 - 0.78 \text{ kcal mole}^{-1}$$
$$= 3.53 \text{ kcal mole}^{-1}$$

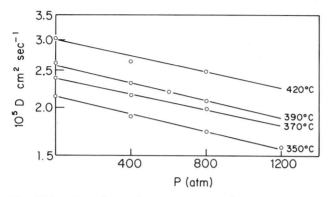

Fig. 6.7.3. Plots of D against pressure at various temperatures for Na^{22} diffusion in fused NaNO$_3$.

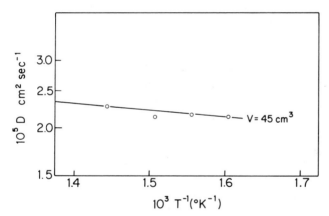

Fig. 6.7.4. Plot of D (at constant volume) against T^{-1} for Na^{22} diffusion in fused NaNO$_3$.

and

$$E_j^* = 0.78 \text{ kcal mole}^{-1}$$

The expectation of the density fluctuation hole model that the standard enthalpy of hole formation should be large compared to the activation energy of jumping is thus borne out by experiment. Since the microjump model assumes that no holes comparable in size to the particle are present, the expectation from this model would be that $E_j^* \approx E_p$.

9 The activation volume for any rate process is given by

$$\Delta V^* = [\partial(\Delta G^*)/\partial P]_T \tag{6.9.3}$$

where ΔG^* is the free energy of activation for that process. Since diffusion is a rate process, the diffusion coefficient D can be written in the form

$$D = D_0 \exp(-\Delta G^*/RT) \tag{6.9.4}$$

where D_0 is independent of pressure and temperature. Substituting (6.9.4) into (6.9.3) yields

$$\Delta V^* = -RT[\partial(\ln D)/\partial P]_T \tag{6.9.1}$$

On the basis of the density fluctuation hole model, we may consider that

$$\Delta G^* = \Delta G_h^\circ + \Delta G_j^* \tag{6.9.5}$$

where ΔG_h° is the standard free energy of hole formation and ΔG_j^* the free energy of activation for jumping. Differentiating (6.9.5) with respect to pressure at constant temperature yields

$$\Delta V^* = \Delta V_h + \Delta V_j^* \tag{6.9.6}$$

Now, $\Delta V_h = V_h$, the most probable molar hole volume, since the volume of a hole before it is formed is zero. Consequently, (6.9.6) becomes (6.9.2):

$$\Delta V^* = V_h + \Delta V_j^*$$

Equation (6.9.1) shows that ΔV^* can be obtained from the slope of the plot of log D against P at constant temperature (Fig. 6.7.3 for the diffusion of Na^{22} in molten $NaNO_3$). From the slope of the plots at 390°C, we have

$$2.303[(d \log D)/dP]_T = -1.95 \times 10^{-4} \text{ atm}^{-1}$$

and from (6.9.1),

$$\Delta V^* = 82.06 \times 663 \times 1.95 \times 10^{-4} \text{ cm}^3 \text{ mole}^{-1}$$
$$= 10.6 \text{ cm}^3 \text{ mole}^{-1}$$

For ΔV^* to be expressed in cm³, the gas constant R must be expressed in cm³ atm deg⁻¹ mole⁻¹.

Similarly for the diffusion of ¹³⁴Cs in $NaNO_3$ at 390°, we have, by plotting the data of Table 6.9.1 (Fig. 6.9.1),

$$\Delta V^* = 82.06 \times 663 \times 2.74 \times 10^{-4} \text{ cm}^3 \text{ mole}^{-1}$$
$$= 14.9 \text{ cm}^3 \text{ mole}^{-1}$$

From Eq. (6.1.3), the molar mean hole volume V_h is given by

$$V_h = 0.68 N_A (kT/\gamma)^{3/2} \qquad (6.9.7)$$

where N_A is Avogadro's number. For fused sodium nitrate at 390°C, we have

$$V_h = 0.68 \times 6.023 \times 10^{23} (1.38 \times 10^{-16} \times 663/110)^{3/2}$$
$$= 9.8 \text{ cm}^3 \text{ mole}^{-1}$$

Substituting the values of ΔV^* for Na^+ and Cs^+ in fused $NaNO_3$ and the value of V_h into (6.9.2) permits the values of ΔV_j^* to be calculated (Table 6.9.2).

For Na^+ diffusion, the fact that the magnitude of the volume change associated with the diffusion process ΔV^* is approximately equal to the most probable hole volume is a clear and direct demonstration of the importance of the part played by holes in diffusive

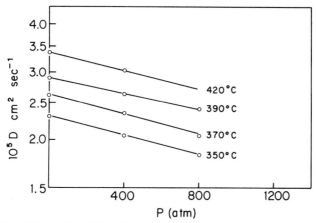

Fig. 6.9.1. Plot of D against pressure at constant temperature for ¹³⁴Cs diffusion in fused $NaNO_3$.

**TABLE 6.9.2.   Activation Volumes
for Diffusion Processes in Fused $NaNO_3$**

	Na^+	Cs^+
ΔV^*, cm³ mole⁻¹	10.6	14.9
V_h, cm³ mole⁻¹	9.8	9.8
ΔV_j^*, cm³ mole⁻¹	0.9	5.1

transport. The fact that ΔV_j^* is close to zero for Na^+ and smaller alkali cations is in direct contradiction to a microjump model. For the diffusion of Cs^+, ΔV_j^* is not close to zero. This is to be expected on the basis of the density fluctuation hole model for the diffusion of an ion with a volume larger than the mean hole volume. The crystallographic radius of Cs^+ is 1.69 Å.

CHAPTER 7

THE ELECTRICAL DOUBLE LAYER

|1| Calculate the electric field, in esu per square centimeter and also in volts per centimeter, in the double layer at a metal electrode–aqueous solution interface when the excess surface charge is 10 and 40 μC cm^{-2} (microcoulombs per square centimeter). Assume there is no specific adsorption but that the solution is concentrated.

|2| The viscosity of water is 0.01 P at 25°C. Calculate the velocity of a colloidal particle in a field of 1 V cm^{-1} if its zeta potential is 13 mV. $\epsilon_w = 78.3$.

|3| Calculate the charge density on the metal as a function of potential and then the capacitance as a function of potential from the electrocapillary data of Table 7.3.1 (see page 86).

|4| Table 7.4.1 (see page 86) lists the interfacial tension as a function of potential and solution concentration for the mercury–lithium chloride solution interface. Table 7.4.2 (see page 87) lists the charge on the mercury electrode under the same conditions, and molar activity coefficients are listed for various concentrations of lithium chloride in Table 7.4.3 (see page 87). All potentials are referred to that of a reversible calomel electrode in the same solution, which is reversible to chloride ions. Calculate the cationic and anionic Gibbs surface excesses for each potential and solution concentration.

TABLE 7.3.1. Variation of Interfacial Tension
of a Mercury–1.0 M CsCl Interface with Potential Difference*

Potential difference, mV vs reversible calomel electrode	Interfacial tension, dynes cm^{-1}
−0	345.0
−100	376.4
−200	397.1
−300	410.5
−400	418.7
−500	422.6
−600	422.9
−700	419.9
−800	414.0
−900	405.6
−1000	395.1
−1100	382.9
−1200	369.2
−1300	353.6

* Data from Kovac, 1964.

TABLE 7.4.1. Interfacial Tension at Various Potentials
and Concentrations of Lithium Chloride*

Potential, mV vs R.C.E.	Interfacial tension, dynes cm^{-1}					
	3.0 M	1.0 M	0.30 M	0.10 M	0.03 M	0.01 M
−100	372.7	379.2	383.6	385.7	383.1	382.9
−200	391.2	397.6	401.1	402.5	398.7	397.9
−300	405.2	410.8	413.4	414.3	410.6	409.4
−400	414.9	419.3	421.1	421.8	419.0	417.6
−500	420.4	423.5	424.8	425.6	424.1	423.1
−600	421.8	423.9	425.2	426.1	426.3	426.0
−700	419.4	421.0	422.5	424.0	425.6	426.5
−800	413.6	415.2	417.3	419.5	422.4	424.7
−900	405.1	407.0	409.9	413.1	417.0	420.7
−1000	394.3	396.9	400.6	405.0	409.6	414.8
−1100	381.8	385.3	389.8	395.4	400.4	406.9
−1200	367.0	372.2	377.7	384.3	389.7	397.4
−1300	352.7	357.8	364.1	371.9	377.7	386.3
−1400	335.1	341.6	348.8	357.9	364.3	373.8

* Data from Kovac, 1964.

TABLE 7.4.2.   Electrode Charge Density at Various Potentials and Concentrations of Lithium Chloride*

Potential, mV *vs* R.C.E.	Charge density, μC cm^{-2}					
	3.0 *M*	1.0 *M*	0.3 *M*	0.1 *M*	0.03 *M*	0.01 *M*
−100	20.8	21.3	20.5	19.6	17.6	17.1
−200	16.2	15.8	14.7	14.1	13.7	13.1
−300	11.9	10.8	9.9	9.5	10.1	9.8
−400	7.6	6.3	5.6	5.5	6.7	6.8
−500	3.4	2.2	2.0	2.1	3.6	4.2
−600	−0.5	−1.3	−1.2	−0.8	0.7	1.7
−700	−4.1	−4.5	−4.0	−3.3	−1.9	−0.7
−800	−7.2	−7.1	−6.4	−5.4	−4.4	−2.9
−900	−9.7	−9.2	−8.4	−7.3	−6.5	−5.0
−1000	−11.6	−11.0	−10.1	−8.9	−8.4	−6.9
−1100	−13.0	−12.4	−11.5	−10.4	−9.9	−8.7
−1200	−14.3	−13.4	−12.8	−11.8	−11.3	−10.3
−1300	−15.9	−15.3	−14.3	−13.2	−12.7	−11.8
−1400	−18.7	−17.4	−16.4	−14.8	−14.1	−13.2

* Data from Kovac, 1964.

TABLE 7.4.3.   Mean Ion Activity Coefficients in Molar Units for Lithium Chloride Solutions

C_{LiCl}, *M*	$y_{\pm}$
0.010	0.89
0.030	0.84
0.10	0.78
0.30	0.72
1.0	0.72
3.0	0.96

5   The data listed in Table 7.5.1 were obtained from electrocapillary measurements on the mercury–0.1 *M* sodium thiocyanate solution interface. Calculate the amount of contact-absorbed CNS$^-$, expressed in units of charge, at each potential. Construct a plot of contact adsorbed thiocyanate against charge on the electrode. Comment on the result.

TABLE 7.5.1.   Cationic and Anionic Surface Excesses
for 0.1 M Sodium Thiocyanate Solutions*

Potential, mV	$F\Gamma_+$, μC cm^{-2}	$F\Gamma_-$, μC cm^{-2}
−300	8.2	−28.5
−400	8.7	−23.3
−500	8.5	−17.9
−600	8.2	−13.0
−700	8.0	−8.7
−800	8.1	−5.3
−900	8.5	−2.7
−1000	9.3	−0.8
−1100	10.4	0.4
−1200	11.9	1.0

* Data from Kovac, 1964.

6   Express the Volta potential difference in terms of the total charge on the metal electrode.

Consider water dipoles attached to an electrode surface. Suppose that they (a) fully occupy the surface; (b) can have only two positions: these are symmetric and can be designated to as ↓ and ↑; and (c) are in equilibrium with water molecules in solution.

Show that the potential difference at the interface due to the water molecules is given by

$$\Delta\chi = \frac{4\pi\mu N_t}{\epsilon} \tanh\left[\frac{\mu X}{kT} - \frac{Uc}{kT}(\theta\downarrow - \theta\uparrow)\right] \qquad (7.6.1)$$

where N_t is the total number of water molecules in the electrode surface, ϵ is the dielectric constant of the oriented water layer, μ is the dipole moment, X is the electric field strength due to the electrode, U is the interaction between two neighboring dipoles, c is the coordination number, and $\theta\uparrow$ and $\theta\downarrow$ are the fractions of the two types of water in this model.

Plot $\Delta\chi$ against the electrode charge q_M over the range 0 to +20 μC cm^{-2}. Pay particular attention to N_t; the value to be used should be consistent with a fairly small $\Delta\chi$ (e.g., ±100 mV) at $q_M = $ ±20 μC cm^{-2}. Note that since $U = f(N_t)$, there is feedback between the U values chosen and N_t. What average distance between water molecules on the surface is consistent with $\Delta\chi < 100$ mV ?

| 7 | Consider the interaction energy between anions in contact with an electrode interface, taking into account the image |

Consider the interaction energy between anions in contact with an electrode interface, taking into account the image energy of the ions with the electrode. Show that the total lateral interaction energy is given by

$$W_{LI} = \frac{\pi^{7/2} e_0^2 r_i^2 n_{CA}^{3/2}}{4\epsilon} \left(1 - \frac{\pi^3 r_i^2 n_{CA}}{20}\right) \qquad (7.7.1)$$

where r_i is the ionic radius, n_{CA} is the number of contact-adsorbed ions per cm² on the surface, and ϵ is the dielectric constant in the first layer.

| 8 | Prove that if contact-adsorbed ions interact coulombically with the electrode of charge density q_M, then, at fractional coverage θ, the isotherm is

$$n_{CA} = (1 - \theta)\frac{2N_A r_i a_i}{1000} \exp\left[-\frac{\Delta G_C^\circ}{RT} + \frac{4\pi z_i e_0 q_M(x_1 - x_2)}{\epsilon kT}\right.$$

$$\left. - \frac{\pi^{7/2} e_0^2 r_i^2 n_{CA}^{3/2}}{4\epsilon kT}\left(1 - \frac{\pi^3 r_i^2 n_{CA}}{20}\right)\right] \qquad (7.8.1)$$

where r_i, z_i, and a_i are the radius, charge, and activity in the bulk of the solution of the contact-adsorbed ion, ΔG_C° is the chemical component of the standard free energy change associated with the contact adsorption of the ions, and x_1 and x_2 are the distances from the metal surface to the inner Helmholtz plane (IHP) and the outer Helmholtz plane (OHP), respectively.

| 9 | Show that the isotherm based on Coulombic repulsion, (7.8.1), gives a relation between the contact adsorbed charge q_{CA} and the charge on the metal q_M which passes through a point of inflection. Further, this point of inflection is given in terms of θ, the fraction of the inner Helmholtz plane (IHP) occupied by contact adsorbed ions, by

$$\theta = (4/3B)^{2/3} \qquad (7.9.1)$$

where

$$B = \frac{\pi^{7/2} e_0^2 r_i^2}{4\epsilon kT}\left(\frac{q_{max}N_A}{F}\right)^{3/2} \qquad (7.9.2)$$

and r_i is the ionic radius, ϵ is the dielectric constant in the IHP, N_A is Avogadro's number, and q_{max} is the charge density associated with complete coverage by contact-adsorbed ions.

Table 7.9.1 lists the contact-adsorbed charge, for a number of anions, corresponding to the electrode potential at which the hump is observed on the capacitance–potential curve. These values of the contact adsorbed charge $q_{CA-hump}$ were experimentally determined on mercury. Calculate the values of q_{CA} at the inflection of the q_{CA} versus q_M curve and compare them to $q_{CA-hump}$. Discuss the results.

TABLE 7.9.1.   Contact-Adsorbed Charge at the Capacitance Hump $q_{CA-hump}$ for Various Anions of Radius r_i

Anion	Concentration, M	r_i, Å	$q_{CA-hump}$, $\mu C\ cm^{-2}$
I^-	0.25	2.00	−18.3
Cl^-	3.0	1.81	−11.1
CN^-	0.3	1.95	−6.7
BrO_3^-	0.3	1.6	−13.2
ClO_4^-	0.3	2.5	−6.7

10 A metal electrode M is placed in a solution in experiment A, and in experiment B, the metal is placed immediately outside the solution. In each case, the electrode M is connected to another electrode in solution that is reversible to metal ions in the solution. The potential between the electrodes is adjusted for zero current and the potential measured as 0.389 and 0.452 V for experiments A and B, respectively.

Deduce equations which permit the calculation of the metal–solution Volta potential difference. What is its numerical value in this case?

11 A platinum wire passes through a platinum ring. The wire is strung tautly between two fixtures and the whole is immersed in solution. The potential of the wire and ring can be varied by conventional electrochemical circuits and measured against a reference electrode in the solution. By gradually tilting the wire from the horizontal, the coefficient of sliding friction can be obtained.

Show that the coefficient of friction will vary with potential in a parabolic way and that it passes through a maximum at the potential of zero charge. Plot, numerically, the coefficient of friction against electrode charge over the range $q_M = -10$ to $+10$ μC cm^{-2}.

$\boxed{12}$ Consider a model of organic adsorption on electrodes in which the following assumptions are made:

(a) There is equilibrium between the adsorbed organic material and the electrode.

(b) The organic does not dissociate as it transfers from the electrode to the solution and vice versa.

(c) Only electrostatic interactions between the electrode and (i) water molecules, (ii) organic dipoles, are significant.

On this basis and with the modelistic assumption that water exists either in the $\uparrow$ or $\downarrow$ position, show that (for an organic without electrical interaction with the surface)

$$\frac{\theta_{\text{org}}}{1 - \theta_{\text{org}}} = C_{\text{org}} A \exp\left[-\frac{nY}{kT}\left(\frac{4\pi\mu q_M}{\epsilon} - YUc\right)\right] \quad (7.12.1)$$

where A is a constant for a given system and temperature, n is the number of water molecules displaced by one organic molecule, μ is the dipole moment of water, U is the lateral energy of interaction between water molecules, c is the coordination number of water molecules on the surface, and Y is given by

$$Y = \theta\uparrow - \theta\downarrow$$

Calculate and plot θ_{org} as a function of q_M for $A = 10$.

Examine the effect on this equation of taking into account the electrical interaction of the organic molecule with the electrode.

By considering the change in the double-layer energy that results from the interaction of large organic molecules with the electrode, it can be shown that θ_{org} should be decreased as the potential is changed in either direction from the p.z.c. (Frumkin, 1926; Frumkin and Damaskin, 1964). Make numerical calculations to compare the relative importance of the effect with that calculated on the basis of the model presented in the first part of this problem.

ANSWERS

1 In concentrated solutions, the thickness of the diffuse double layer become negligible and, in the absence of specific adsorption, the double layer approximates a parallel plate capacitor with charge equal to the surface charge on the electrode. For such a capacitor, the electric field X is related to the charge q by

$$X = 4\pi q/\epsilon \qquad (7.1.1)$$

where ϵ is the dielectric constant of the material between the electrode and the outer Helmholtz plane (OHP). ϵ is certainly not that of bulk water. For the totally restricted water molecules adsorbed on the surface, $\epsilon \approx 6$, but its value increases rapidly with distance from the electrode to the value of bulk water. Taking a value of, say, 20 for ϵ and substituting numerical values into (7.1.1) yields

$$X = 4\pi \times 10^{-5}/20 \quad \text{coulomb cm}^{-2}$$
$$= 4\pi \times 10^{-5} \times 3 \times 10^{9}/20 \quad \text{esu cm}^{-2}$$
$$= 1.9 \times 10^{4} \text{ esu cm}^{-2}$$
$$= 1.9 \times 10^{4} \times 300 \quad \text{V cm}^{-1}$$
$$= 5.6 \times 10^{6} \text{ V cm}^{-1}$$

when

$$q = 40 \ \mu\text{C cm}^{-2}$$
$$X = 7.5 \times 10^{4} \text{ esu cm}^{-2}$$
$$= 2.3 \times 10^{7} \text{ V cm}^{-1}$$

3 The charge density q_M on the mercury electrode is given by the Lippmann equation:

$$(\partial\gamma/\partial V)_\mu = -q_M \qquad (7.3.1)$$

where γ is the interfacial tension, V the potential, and μ the chemical potential of species in the solution. For γ in dynes per centimeter and V in volts, the units of q_M are dynes per centimeter per volt. The conver-

sion to coulombs per square centimeter can be derived in a number of
ways:

$$dyne/cm\ V = dyne\ A/cm\ VA$$
$$= dyne\ A\ sec/cm\ J$$
$$= 10^{-7}\ A\ sec\ cm^{-2}$$
$$= 10^{-7}\ coulomb\ cm^{-2}$$

Figure 7.3.1 shows a plot of the data of Table 7.3.1 and the evaluation of q_M from the slope of the plot. The charge density on the metal
is plotted against potential in Fig. 7.3.2. The differential capacity C
is given by

$$C = dq_M/dV \qquad (7.3.2)$$

and is the tangent to the curve of q_M against V (Fig. 7.3.2). At
-600 mV RCE, the differential capacitance is seen to be 32.5 μF cm^{-2}.
Capacitances at other potentials are similarly calculated.

[5]   A good deal of evidence exists to show that cations, with the
exception of large cations such as Cs^+, $N(CH_3)_4^+$ etc., are not
specifically adsorbed. Consequently, the entire cationic excess charge
can be considered to be cationic diffuse layer charge q_d^+. From Gouy–
Chapman theory, the cationic charge in the diffuse layer is related to
the potential at the OHP, ψ_0, by

$$q_d^+ = (\epsilon nRT/2\pi)^{1/2}[\exp(-z_+F\psi_0/2RT) - 1] \qquad (7.5.1)$$

Hence, by equating q_d^+ to $z_+F\Gamma_+$, where Γ_+ is the experimentally
determined Gibbs surface excess of the cation, we can calculate ψ_0
from (7.5.1). Knowing ψ_0, the anionic charge in the diffuse layer q_d^-
can be calculated from

$$q_d^- = -(\epsilon nRT/2\pi)^{1/2}[\exp(-z_-F\psi_0/2RT) - 1] \qquad (7.5.2)$$

For the charge to have units of coulombs per square centimeter,
the term on the right of (7.5.1) and (7.5.2) must have these units. For n,
the bulk concentration of the ion, in moles per cubic centimeter and R

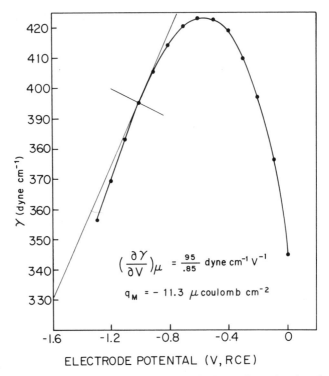

Fig. 7.3.1. Evaluation of the electrode charge from the plot of surface tension against potential for the mercury–0.1 M CsCl interface. $(\partial\gamma/\partial V)_\mu = 95/0.85$ dynes cm^{-1} V^{-1}; $q_M = -11.3$ μC cm^{-2}.

in ergs per mole per degree, the units can be converted in the following way:

$$
\begin{aligned}
(\text{erg/cm}^3)^{1/2} &= (10^{-7})^{1/2}(\text{VA sec/cm}^3)^{1/2} \\
&= (10^{-7}/300)^{1/2}(\text{statvolt coulomb/cm}^3)^{1/2} \\
&= (10^{-7}/300)^{1/2}(\text{esu coulomb/cm}^4)^{1/2} \\
&= (10^{-7}/3 \times 10^9 \times 300)^{1/2}(\text{coulomb/cm}^4)^{1/2} \\
&= 0.333 \times 10^{-9} \text{ coulomb cm}^{-2}
\end{aligned}
$$

Evaluating the term $(\epsilon nRT/2\pi)^{1/2}$ for a 0.1 M solution of a 1:1 electrolyte, we have

$$
\begin{aligned}
\left(\frac{\epsilon nRT}{2\pi}\right)^{1/2} &= 0.333 \times 10^{-9}\left[\frac{78.5 \times 10^{-4} \times 8.314 \times 10^7 \times 298}{2\pi}\right]^{1/2} \\
&= 1.85 \times 10^{-6} \text{ coulomb cm}^{-2}
\end{aligned}
$$

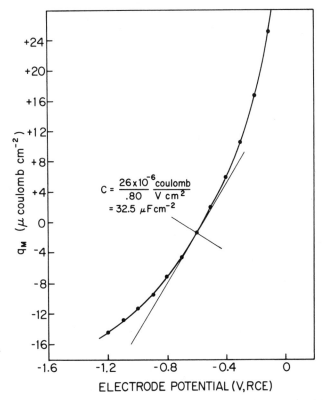

Fig. 7.3.2. Electrode charge plotted against electrode potential for the mercury–0.1 M CsCl interface. $C = (26 \times 10^{-6}/0.80)$ coulomb/V-cm² = 32.5 μF cm⁻².

The potential in the OHP, ψ_0, can now be evaluated from (7.5.1) for the case when the cationic excess is 8.2 μC cm⁻² (see the first line of Table 7.5.1):

$$\psi_0 = -(2RT/z_+F)\ln[(8.2/1.85) + 1]$$
$$= -(2 \times 8.314 \times 298/96,500)\ln(5.43)$$
$$= -0.086 \text{ V}$$

Substituting for ψ_0 in (7.5.2) yields

$$q_{d^-} = -1.85 \times 10^{-6}[\exp(0.086 \times 96,500/2 \times 8.314 \times 298) - 1]$$
$$= +1.50 \ \mu\text{C cm}^{-2}$$

The positive sign indicates a deficiency of anions in the diffuse double layer. Since the contact-adsorbed charge q_{CA} is related to the total anionic excess charge by

$$F\Gamma_- = q_{CA} + q_d^-$$ (7.5.3)

then

$$q_{CA} = -28.5 - 1.5$$
$$= -30.0 \ \mu C \ cm^{-2}$$

The value of the electrode charge q_M corresponding to this value of the contact-adsorbed charge is computed from

$$-q_M = F\Gamma_- + F\Gamma_+$$ (7.5.4)

and from Table 7.5.1,

$$-q_M = -28.5 + 8.2$$
$$= -20.3 \ \mu C \ cm^{-2}$$

Using the data of Table 7.5.1, values of q_{CA} are computed for the entire range of values of q_M. The results are plotted in Fig. 7.5.1. Careful examination of Fig. 7.5.1 shows that the plot passes through a point of inflection. The point of inflection is a characteristic feature of such plots and consequently it is essential that any isotherm derived for contact adsorption on electrodes predict the existence of the inflection point. In Problems 7–9 of this chapter, an isotherm based on Coulombic repulsion, taking into account image charges, will be developed and tested. Further, it will be shown that the point of inflection corresponds to the "capacitance hump" exhibited by capacitance–electrode potential plots.

| 7 | Let us divide a unit area in the IHP into as many cells as there are contact-adsorbed anions, and make the assumption that the cell centers, which correspond to the time average positions of the ions, are arranged according to a hexagonal pattern. Choosing one ion as the reference ion, there will be a ring containing six ions around the reference ion (Fig. 7.7.1). The nth ring of anions contains $6n$ ions and, if we consider only univalent anions, it will have a total charge of $-6ne_0$.

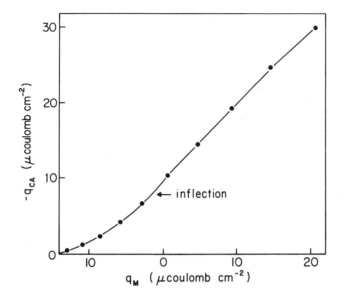

Fig. 7.5.1. Plot of contact-adsorbed charge against electrode charge showing the point of inflection for the mercury–0.1 M NaCNS interface.

As a mathematical convenience, we will neglect the discreteness of the charges in the ring and consider it to be a continuum of charge with a charge density per radian, J_n , given by

$$J_n = -6ne_0/2\pi \tag{7.7.2}$$

The potential ψ at the reference ion due to the charge on the nth ring is

$$\psi = 2\pi J_n/\epsilon nr \tag{7.7.3}$$

where r is the distance between the centers of the cells (Fig. 7.7.1). Consequently, the interaction energy between the reference anion and the charge of the nth ring, $-\psi e_0$ (which is repulsive and thus of positive sign), is given by

$$-\psi e_0 = -2\pi J_n e_0/\epsilon nr \tag{7.7.4}$$

Consider now the interaction energy between the reference anion and the nth ring of image charge (Fig. 7.7.2). If the electrode-to-

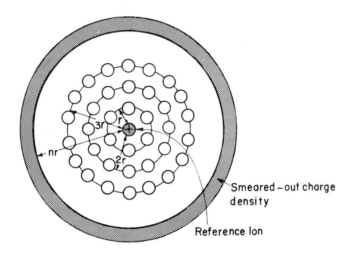

Fig. 7.7.1.   Contact-adsorbed ions arranged in a hexagonal pattern. Here, nr in the distance between the reference ion and the nth ring of charge (shown as a continuum).

reference-ion distance is r_i, then the distance between the reference ion and the ring of image charge is $[(nr)^2 + (2r_i)^2]^{1/2}$. Consequently, this interaction energy, $-e_0\psi_{image}$ (which is attractive and hence of negative sign), is given by

$$-e_0\psi_{image} = 2\pi J_n e_0/\epsilon(n^2 r^2 + 4r_i^2)^{1/2} \qquad (7.7.5)$$

Hence the interaction energy between the reference anion and the nth ring of charge and its image is given by the sum of (7.7.4) and (7.7.5), namely

$$(2\pi \mid J_n \mid e_0/\epsilon nr)\{1 - [1 + (2r_i/nr)^2]^{-1/2}\} \qquad (7.7.6)$$

The total lateral interaction energy W_{LI} is therefore the sum of the interactions due to all the rings and is given by

$$W_{LI} = \sum_{n=1}^{\infty} (2\pi \mid J_n \mid e_0/\epsilon nr)\{1 - [1 + (2r_i/nr)^2]^{-1/2}\} \qquad (7.7.7)$$

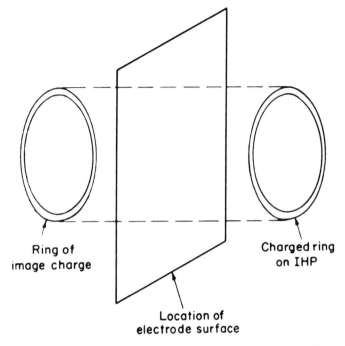

Ring of
image charge

Charged ring
on IHP

Location of
electrode surface

Fig. 7.7.2.   The image of the contact-adsorbed charge is induced behind
the electrode surface. The sign of the image charge is opposite to that of
the contact-adsorbed charge but equal in absolute value.

Since $(2r_i/nr)^2$ is less than unity, $[1 + (2r_i/nr)^2]^{-1/2}$ may be replaced
by its binomial expansion:

$$[1 + (2r_i/nr)^2]^{-1/2} = 1 - \tfrac{1}{2}(2r_i/nr)^2 + \tfrac{3}{8}(2r_i/nr)^4 \qquad (7.7.8)$$

where higher-order terms are neglected. Substituting (7.7.8) into (7.7.7)
yields

$$W_{LI} = \sum_{n=1}^{\infty} (4\pi \mid J_n \mid e_0 r_i{}^2/\epsilon r^3 n^3)[1 - (3r_i{}^2/n^2 r^2)] \qquad (7.7.9)$$

Since r is the distance between cell centers (i.e., the ion's diameter),
each cell has an area $\pi(r/2)^2$ and hence there are $4/\pi r^2$ ions per unit
area. However, n_{CA} is the number of contact-adsorbed ions per unit
area, and consequently,

$$r = (4/\pi n_{CA})^{1/2} \qquad (7.7.10)$$

Substituting (7.7.2) and (7.7.10) into (7.7.9) yields

$$W_{LI} = \sum_{n=1}^{\infty} \frac{3r_i^2 e_0^2 \pi^{3/2} n_{CA}^{3/2}}{2\epsilon n^2} \left\{ 1 - \frac{3\pi r_i^2 n_{CA}}{4} \frac{1}{n^2} \right\} \qquad (7.7.11)$$

Inspection of (7.7.11) shows that it contains two summations:

$$\sum_{n=1}^{\infty} n^{-2} = \pi^2/6 \qquad (7.7.12)$$

and

$$\sum_{n=1}^{\infty} n^{-4} = \pi^4/90 \qquad (7.7.13)$$

Substituting (7.7.12) and (7.7.13) into (7.7.11) yields the desired expression (7.7.1):

$$W_{LI} = \frac{\pi^{7/2} e_0^2 r_i^2 n_{CA}^{3/2}}{4\epsilon} \left(1 - \frac{\pi^3 r_i^2 n_{CA}}{20} \right)$$

9   The number of contact-adsorbed ions per square centimeter in the IHP, n_{CA}, is related to the total density of sites for contact adsorption n_T and the fractional coverage θ by

$$n_{CA} = \theta n_T \qquad (7.9.3)$$
$$= \theta n_T e_e N_A / F \qquad (7.9.4)$$
$$= \theta \, | \, q_{max} \, | \, N_A / F \qquad (7.9.5)$$

Introducing (7.9.5) into (7.8.1), and since

$$1 - (\pi^3 r_i^2 n_{CA}/20) \approx 1 \qquad (7.9.6)$$

Eq. (7.8.1) can be written

$$\ln[\theta/(1 - \theta)] = \text{const} + \ln a_i + Aq_M - B\theta^{3/2} \qquad (7.9.7)$$

where

$$\text{const} = \ln(2r_i F/1000 \, | \, q_{max} \, |) - (\Delta G_c^{\circ}/RT) \qquad (7.9.8)$$
$$A = 4\pi e_0(x_2 - x_1)/\epsilon kT \qquad (7.9.9)$$
$$B = (\pi^{7/2} e_0^2 r_i^2/4\epsilon kT)(| \, q_{max} \, | \, N_A/F)^{3/2} \qquad (7.9.10)$$

Equation (7.9.7) will pass through points of inflection where the second derivative $d^2\theta/dq_M{}^2$ equals zero. The differentiation may be performed as follows:

$$A\frac{dq_M}{d\theta} = \frac{1}{\theta} + \frac{1}{1-\theta} + \frac{3}{2}B\theta^{1/2} \qquad (7.9.11)$$

and inverting,

$$d\theta/dq_M = A/P \qquad (7.9.12)$$

where

$$P = [1/\theta(1-\theta)] + \tfrac{3}{2}B\theta^{1/2} \qquad (7.9.13)$$

Now,

$$\frac{d^2\theta}{dq_M{}^2} = \frac{1}{2}\frac{d}{d\theta}\left(\frac{d\theta}{dq_M}\right)^2 \qquad (7.9.14)$$

$$= \frac{1}{2}\frac{d}{d\theta}\left(\frac{A}{P}\right)^2 \qquad (7.9.15)$$

$$= -\frac{A^2}{P^3}\frac{dP}{d\theta} \qquad (7.9.16)$$

where

$$\frac{dP}{d\theta} = \frac{-1 + 2\theta + \tfrac{3}{4}B\theta^{-1/2}\theta^2(1-\theta)^2}{\theta^2(1-\theta)^2} \qquad (7.9.17)$$

$$\approx \frac{-1 + 2\theta + \tfrac{3}{4}B\theta^{3/2} - \tfrac{3}{2}\theta^{5/2}}{\theta^2(1-\theta)^2} \qquad (7.9.18)$$

if the term in $\theta^{7/2}$ is neglected as small (see below). Substituting (7.9.18) into (7.9.16) yields

$$\frac{d^2\theta}{dq_M{}^2} = -\frac{A^2}{2P^3\theta^2(1-\theta)^2}\left(-2 + 4\theta + \frac{3}{2}B\theta^{3/2} - 3B\theta^{5/2}\right) \qquad (7.9.19)$$

The mathematical condition for the inflection point is satisfied when (7.9.19) equals zero. That is, when

$$2 - 4\theta - \tfrac{3}{2}B\theta^{3/2} + 3B\theta^{5/2} = 0 \qquad (7.9.20)$$

Equation (7.9.20) can be factorized as follows:

$$(1 - 2\theta)(1 - \tfrac{3}{4}B\theta^{3/2}) = 0 \qquad (7.9.21)$$

and consequently has two roots θ_1 and θ_2 given by

$$\theta_1 = (4/3B)^{2/3} \tag{7.9.22}$$

$$\theta_2 = 1/2 \tag{7.9.23}$$

The first root θ_1 is smaller than θ_2, and for the first root the error introduced by neglecting $\theta^{7/2}$ [see Eqs. (7.9.17) and (7.9.18)] is negligibly small.

For contact-adsorbed anions, the charge density corresponding to the inflection point, $q_{CA-infl}$, is given by

$$q_{CA-infl} = -\theta_1 n_T e_0 \tag{7.9.24}$$

$$= -(4n_T^{3/2}/3B)^{2/3} e_0 \tag{7.9.25}$$

Substituting (7.9.10) into (7.9.25) and taking into account that

$$n_T = |q_{max}| N_A/F \tag{7.9.26}$$

we have

$$q_{CA-infl} = (4/r_i^{4/3})(4\epsilon^2 k^2 T^2/9\pi^7 e_0)^{1/3} \tag{7.9.27}$$

The dielectric constant ϵ in (7.9.27) refers to the IHP value. The dielectric constant in the double layer increases from 6 in the saturated water layer on the electrode to some intermediate value between 6 and 80 in the OHP. Since contact-adsorbed ions extend a little further from the electrode than the water molecules, a value a little larger than 6 is appropriate for the calculation of $q_{CA-infl}$. The following numerical calculations are based on $\epsilon = 10$.

Substituting numerical values in (7.9.27), we have

$$q_{CA-infl} = -\frac{4}{r_i^{4/3}(10)^{32/3}}\left(\frac{4 \times 10^2 \times (1.38)^2 \times 10^{-32} \times (298)^2}{9\pi^7 \times 4.8 \times 10^{-10}}\right)^{1/3}$$

$$= -6.93 \times 10^4/r_i^{4/3} \quad \text{esu cm}^{-2}$$

$$= -23.1/r_i^{4/3} \quad \mu C \text{ cm}^{-2}$$

when r_i is expressed in angstroms.

Table 7.9.2 lists values of $q_{CA-infl}$ calculated from the ionic radii in Table 7.9.1 using Eq. (7.9.27). Also listed for comparison in Table 7.9.2 are the values of $q_{CA-hump}$, i.e., the experimentally determined contact-adsorbed charge corresponding to the hump on the capacitance–

TABLE 7.9.2.   Comparison of $q_{CA-\text{infl}}$ and $q_{CA-\text{hump}}$ for Various Anions Contact-Adsorbed on Mercury

Ion	$q_{CA-\text{infl}}$, ($\mu C\ cm^{-2}$)	$q_{CA-\text{hump}}$, ($\mu C\ cm^{-2}$)
I^-	-8.9	-18.3
Cl^-	-10.5	-11.4
CN^-	-9.5	-6.7
BrO_3^-	-12.5	-13.2
ClO_4^-	-6.8	-6.7

electrode potential curve. In all cases except iodide, there is good agreement between $q_{CA-\text{infl}}$ and $q_{CA-\text{hump}}$ and this provides strong support for the Coulombic repulsion model of contact adsorption and the contention that the capacitance hump is due to contact adsorption. The poor agreement in the case of iodide may, in part, be due to distortion of the large, easily polarized iodide ion in the electric field of the electrode.

CHAPTER 8

BASIC ELECTRODE KINETICS

1 An electrode reaction for which the exchange current density is 5×10^{-7} A cm^{-2} is proceeding at an overpotential $\eta = -0.199$ V under the control of a potentiostat. What is the current density? Using a fast switch, the potentiostat is disconnected from the electrode and a cathodic current density of 2.05×10^{-4} A cm^{-2} imposed on the electrode. What is the new steady-state overpotential and approximately how long would it take to attain it? Take the electrode area as 0.1 cm^2, the double layer capacitance as 50 μF cm^{-2}.

2 An electrode–solution interface has a capacitance of 50 μF cm^{-2}, and a low-amplitude ($\sim$5 mV) ac signal of 1000 Hz is applied to it. If the exchange current density for the reaction is 3 mA cm^{-2}, calculate the Faradaic resistance. Also work out the Faradaic reactance and impedence (in ohms cm^{-2}).

3 What is Bronsted's law? Suppose that H_2 is being evolved on a cathode at an overpotential of -0.160 V for a rate of 2×10^{-7} mole cm^{-2} sec^{-1} of H_2 at 25°C. The rate-determining step is

$$H_3O^+ + e^- \rightarrow H_{ad} + H_2O \qquad (8.3.1)$$

and the coverage by H_{ad} is assumed to be low. An organic substance is now added to the solution and adsorbs, reducing the average heat of adsorption of H by 12 kcal. Calculate the new rate of the reaction if the overpotential is now -0.222 V at 25°C.

105

> [4] Draw some current–potential lines for both positive and negative overpotentials to illustrate graphically the mechanism of Faradaic rectification. Suppose $\beta = 0.495$; calculate the percentage rectification of a square-wave signal of 200 mV amplitude imposed on the reversible potential ($i_0 = 1 \times 10^{-3}$ A cm^{-2}).

> [5] According to Levich (1970), the energy of activation E^* of a charge-transfer process at an electrode is given by

$$E^* = (E_s + Q)^2/4E_s \qquad (8.5.1)$$

where E_s is the solvent reorganizational energy required to trigger a quantum mechanical transition from the initial to the final state and Q is the heat of reaction. By considering appropriate potential energy curves, show, however, that the relation (8.5.1) does not depend on the nature of the model for charge transfer provided that the energy profiles of the initial and final states can be represented by those of simple harmonic oscillators.

> [6] Using Eq. (8.5.1) and the relation

$$Q = Q_{(\eta=0)} + \eta F \qquad (8.6.1)$$

where η is the overpotential and Q the heat of the reaction, derive an equation for β. Why is such a relation incompatible with experiment? What is the essential point about the consideration expressed in Problem 5, which leads Eq. (8.5.1) into error?

> [7] Do image forces affect electrode-kinetic calculations? Examine this point by comparing the fluctuations in image energy experienced by an ion in the OHP with typical values of energies of activation.

Assuming that the only forces acting on an electron emitted from a metal during a charge-transfer process in the gas phase are the electron metal image interactions and the electron–ion attraction, calculate the potential energy–distance relationship. Then, estimate, roughly, the probability of electron penetration of the barrier in a cathodic reaction. Assume the barrier to be a Gamow type and assume the availability of suitable acceptor states in the ion. Take the electron work function of the metal as 5.0 eV, and the metal–ion distance as 8 Å.

8 In an early theory[†] of homogeneous charge-transfer processes developed by Marcus (1963), the free energy of activation for charge transfer ΔG^* is given by

$$\Delta G^* = \Delta G_o^* + \Delta G_C^* \qquad (8.8.1)$$

where ΔG_C^* is the Coulombic contribution given by Coulomb's law as

$$\Delta G_C^* = z_i z_2 e_0^2 / r \epsilon_{stat} \qquad (8.8.2)$$

for two species, 1 and 2, a distance r apart. ϵ_{stat} is the static dielectric constant of the medium. The organizational contribution ΔG_o^* is given by

$$\Delta G_o^* = \frac{\lambda}{4} + \frac{\Delta G^\circ}{2} + \frac{(\Delta G^\circ)^2}{4\lambda} \qquad (8.8.3)$$

where

$$\lambda = \frac{(ne)^2}{2} \left(\frac{1}{2a_1} + \frac{1}{2a_2} - \frac{1}{r} \right) \left(\frac{1}{\epsilon_{opt}} - \frac{1}{\epsilon_{stat}} \right) \qquad (8.8.4)$$

and ΔG° is the standard free energy charge of the reaction. Here, n is the number of electrons transferred in the reaction, a_1 and a_2 are the effective radii of the two reactants, $r = a_1 + a_2$, and ϵ_{opt} is the optical dielectric cunstant of the medium. The effective radius a of a reactant is given approximately by the sum of the radius of the central ion and the diameter of the ligand.

For bimolecular homogeneous reactions, the rate constant k can be written as

$$k = Z \exp(-\Delta G^*/RT) \qquad (8.8.5)$$

and the frequency factor Z has a value of approximately 10^{11} liters mole^{-1} sec^{-1}.

Using Eq. (8.8.5) and the data of Table 8.8.1, calculate experimental ΔG^* values for the processes given in the table. From Eqs.

[†] In later publications, this author took into account the thermal activation of the bonds in inner shells in complex ions, as well as the energy changes concerned with the electrostatic fluctuations of outer water layers.

(8.8.1)–(8.8.4) and the data of Tables 8.8.1 and 8.8.2, calculate the earlier theory values of ΔG^*. Test this theory by plotting the calculated ΔG^* values against the experimental values and discuss the result.

TABLE 8.8.1.   Rate Constants and Standard Free Energy Changes
of Some Homogeneous Charge-Transfer Reactions

Reactants	Temp., °C	$\Delta G°$, kcal mole^{-1}	k, liters mole^{-1} sec^{-1}
$Ce^{3+} + Ce^{4+}$	25	0	1.1×10^{-4}
$Co^{2+} + Co^{3+}$	25	0	0.75
$Co(NH_3)_6^{2+} + Co(NH_3)_6^{3+}$	64.5	0	$< 10^{-8}$
$Cr^{2+} + Cr^{3+}$	24.5	0	2×10^{-5}
$Cr^{2+} + Fe^{3+}$	25	-27.21	$\sim 2 \times 10^{-3}$
$Cr^{2+} + Co(NH_3)_6^{3+}$	25	-11.76	9.0×10^{-5}
$Fe^{2+} + Ce^{4+}$	25	-15.45	7.3×10^{2}
$Fe^{2+} + Co^{3+}$	0	-24.67	10
$Fe^{2+} + Fe^{3+}$	25	0	4.2
$Fe^{2+} + Fe(phen)_3^{3+}$	25	-7.15	3.7×10^{4}
$Fe(CN)_6^{4-} + Ce^{4+}$	25	-17.52	1.9×10^{6}
$Fe(CN)_6^{4-} + Fe(CN)_6^{3-}$	4	0	3.5×10^{2}
$Fe(CN)_6^{4-} + IrCl_6^{2-}$	25	-5.76	3.8×10^{5}
$Fe(CN)_6^{4-} + Mo(CN)_8^{4-}$	25	-2.77	3×10^{4}
$Fe(dipy)_3^{2+} + Ce^{4+}$	25	-8.99	1.96×10^{5}
$Fe(phen)_3^{2+} + Co^{3+}$	25	-17.99	1.6×10^{4}
$Fe(phen)_3^{2+} + Fe(phen)_3^{3+}$	0	0	$> 10^{5}$
$MnO_4^{2-} + MnO_4^{-}$	0	0	7.0×10^{2}
$Mo(CN)_8^{4-} + Ce^{4+}$	25	-14.76	1.4×10^{7}
$Mo(CN)_8^{4-} + IrCl_6^{2-}$	18	-3.0	> 105
$Mo(CN)_8^{4-} + Mo(CN)_8^{3-}$	25	0	3×10^{4}
$V^{2+} + Co(NH_3)_6^{3+}$	25	-11.76	9.0×10^{-5}
$W(CN)_8^{4-} + Ce^{4+}$	25	-20.75	$> 10^{8}$
$W(CN)_6^{4-} + W(CN)_8^{3-}$	25	0	7×10^{4}

9 | The so-called electrostatic model of electrode processes predicts that

$$(k_{hom}/Z_{hom})^{1/2} = (k_{het}/Z_{het}) \tag{8.9.1}$$

TABLE 8.8.2. Radii of Some Central Ions and Diameters of Some Ligands

Central ion	Radius, Å	Ligand	Diameter, Å
Ce^{4+}	0.92	Cl^-	3.62
Co^{2+}	0.72	CN^-	3.84
Co^{3+}	0.62	H_2O	2.76
Cr^{2+}	0.89	NH_3	3.38
Cr^{3+}	0.63	O^{--}	2.64
Fe^{2+}	0.74	O^-	3.52
Fe^{3+}	0.64	Phenanthroline	~5.77
Ir^{4+}	0.68	Dipyridyl	~5.21
Mn^{6+}	~0.50		
Mn^{7+}	0.46		
Mo^{4+}	~0.66		
Mo^{5+}	0.70		
W^{4+}	0.70		
W^{5+}	0.66		

where k_{hom} and k_{het} are the rate constants for exchange reactions of the homogeneous type,

$$Fe^{3+} + Fe^{2+} \rightleftarrows Fe^{2+} + Fe^{3+} \qquad (8.9.2)$$

and the heterogeneous type,

$$Fe^{3+} + e^- \rightleftarrows Fe^{2+} \qquad (8.9.3)$$

respectively; Z_{hom} and Z_{het} are the respective frequency factors. Experimental data obey Eq. (8.9.1), and this was at one time thought to be good evidence in favor of the solvent activation model. Show however, that (8.9.1) holds regardless of the type of activation for models of electrode processes.

10   The expression of time-dependent perturbation theory for the probability of transition between two states i and f is

$$\langle \Psi_f \mid H \mid \Psi_i \rangle \qquad (8.10.1)$$

where H is the Hamiltonian for the transition. State clearly the kind of transition to which this equation is applicable and give two examples outside electrode kinetics.

Calculate the wavelength of an electron which has an energy of 1 eV. Consider the magnitude of this wavelength and then discuss whether it would be justified to apply (8.10.1) to a calculation of the preexponential factor of the rate expression for an interfacial redox reaction.

11. A physical interpretation of the Levich (1970) theory of charge transfer can be made in the following way. Surrounding the ion, water dipoles oscillate at a frequency $\nu_0 \approx 10^{11}$ Hz. These oscillations induce energy into the solvated ion, and when this energy has reached a value such that the energy of an electron level in the ion is that of the Fermi energy of the substrate, a quantum mechanical transition of an electron from the metal to the ion takes place.

Examine this model assuming the necessary energy of activation of the solvated ion is about 1.0 eV. Calculate the energy of an individual oscillator and determine the number of oscillators that would be involved if the ion were to be activated in this way. Approximately how far from the central ion does this number of oscillators extend? Determine the probability that a given central ion is activated 1 eV above its ground energy in a solution as a result of electrostatic fluctuations to which it is subject from the surrounding solvent. Compare this probability with that due to thermal excitation of the rotational–vibrational levels of the ion.

12. Consider hydrogen evolution with a rate-determining step

$$H_3O^+ + e^- \rightarrow H_{ad} + H_2O \qquad (8.12.1)$$

Consider the potential energy–distance relations of the initial and final states to be represented by two intersecting Morse curves. Deduce:

(a) An expression in terms of Morse constants for the symmetry factor β and examine the value of the coefficient $d\beta/d\eta$ over a range of 2 V. Utilize the values: the heat of activation $\Delta H^* = 23$ kcal mole^{-1}; the heat of dissociation of H_3O^+ to H_2O and H^+ (in the gas phase) is 180 kcal mole^{-1}; the heat of dissociation of H_{ad} is 45 kcal mole^{-1}; the enthalpy of reaction (8.12.1) is 55 kcal mole^{-1}; the Morse constant a for H_3O^+ and H_{ad} may be taken as 3×10^8 cm^{-1}; the separation between the zero-point energy states on the reaction coordinate for H_3O^+ and H_{ad} may be taken as 3.3×10^{-8} cm.

(b) The condition for barrierless electron transfer.

ANSWERS

| 1 |

For a simple electrode process, the current density i is given by the Butler–Volmer equation:

$$i = i_0\{\exp[(1 - \beta)\,\eta F/RT] - \exp[-\beta\eta F/RT]\} \qquad (8.1.1)$$

where i_0 is the exchange density and β, the symmetry factor, is always close to 0.5. In the form (8.1.1), i has positive or negative values for anodic or cathodic processes, respectively. The overpotential η is defined by

$$\eta = V - V_R \qquad (8.1.2)$$

where V and V_R are the electrode potential and the reversible potential of the electrode process, respectively.

When $|\eta| > 50$ mV at room temperature, one of the exponential terms in (8.1.1) becomes negligible compared to the other. Substituting given values into (8.1.1) yields

$$i = -5 \times 10^{-7}[\exp(0.5 \times 0.199 \times 96500/8.314 \times 298)]$$

$$= -2.41 \times 10^{-5} \text{ A cm}^{-2}$$

After the imposition of the cathodic current step of 2.05×10^{-4} A cm^{-2}, the steady-state overpotential is given by the solution of

$$-2.05 \times 10^{-4} = -5 \times 10^{-7}[\exp(0.5 \times 96500\eta/8.314 \times 298)]$$

$$\eta = -0.309 \text{ V}$$

Since the anodic contribution to the net current is negligible under the present conditions, the charge transfer resistance R_{CT} may be obtained by differentiating the cathodic part of (8.1.1) with respect to η, thus

$$di/d\eta = (i_0\beta F/RT)\exp(-\beta\eta F/RT) \qquad (8.1.3)$$
$$= i\beta F/RT \qquad (8.1.4)$$

Therefore, if A is the area of the electrode,

$$R_{CT} = (d\eta/di)/A = RT/i\beta FA \qquad (8.1.5)$$

Substituting numerical values into (8.1.5) yields

$$R_{CT} = \frac{8.314 \times 298}{2.05 \times 10^{-4} \times 0.5 \times 96500 \times 0.1}$$

$$= 2505 \text{ ohms}$$

The situation at the electrode is approximately represented by the equivalent circuit in Fig. 8.1.1, and the time constant τ of the circuit to respond to the imposition of a constant current is given by

$$\tau = C_{DL}R_{CT} \qquad (8.1.6)$$

where R_S is the total solution and internal circuit resistance. Substituting numerical values for C_{DL} and R_{CT} yields

$$\tau = 5 \times 10^{-6} \times 2505$$

$$= 12.5 \text{ msec}$$

Since τ is the time required for the system to proceed through $1/e$ of the change, steady-state conditions essentially prevail after 4τ have elapsed; in this case, $\sim$50 msec.

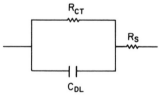

Fig. 8.1.1.   Equivalent circuit for a charge-transfer process at an electrode.

$\boxed{3}$ Bronsted empirically proposed an equation to correlate the specific rate constants k of a series of acid–base reactions with the equilibrium constants K of the reactions. Thus, for reactions of the type

$$A_1H + H_2O \rightleftarrows A_1^- + H_3O^+$$
$$A_2H + H_2O \rightleftarrows A_2^- + H_2O^+$$
$$\cdots \tag{8.3.2}$$
$$A_mH + H_2O \rightleftarrows A_m^- + H_3O^+$$

the Bronsted relation takes the form

$$k = G_A K^\beta \tag{8.3.3}$$

G_A is a constant for the structurally similar series of acids, A_1, A_2,..., A_m, and β is a positive fraction, which is constant for a given reaction series and a given solvent.

Equation (8.3.3) can be written in the form

$$Z \exp(-\Delta G^*/RT) = G_A \exp(-\beta \, \Delta G^\circ/RT) \tag{8.3.4}$$

where Z is the frequency factor, ΔG^* the free energy of activation, and ΔG° the standard free energy of the reaction. Taking logarithms and rearranging (8.3.4) gives

$$-\Delta G^* = RT \ln(G_A/Z) - \beta \, \Delta G^\circ \tag{8.3.5}$$

For the reactions involving the homologous series of acids $A_1, A_2,..., A_m$, G_A and Z are constant and we have

$$\Delta(\Delta G^*) = \beta \, \Delta(\Delta G^\circ) \tag{8.3.6}$$

Thus a change in the standard free energy of the reaction produces a fractional change in the free energy of activation of the reaction. Horiuti and Polanyi (1935) proposed that the same relation should exist for hetrogeneous charge-transfer process at electrodes. Consequently, considering reaction (8.3.1), a change in the heat of adsorption of hydrogen $\Delta(\Delta H)$, will produce a corresponding change $\beta \, \Delta(\Delta H)$ in the free energy of activation of the reaction. The unknown entropy terms are neglected. β is a symmetry factor analogous to the electrochemical symmetry factor and may be taken to have the same value.

The rate of the simple cathodic process (8.3.1) can be given by the Butler–Volmer theory as

$$v_1 = (kT/h)\, C_{H^+}(1 - \theta)\, \exp(-\Delta G_c^*/RT)\exp[-\beta(\Delta\phi)\, F/RT] \quad (8.3.7)$$

where v_1 is the rate, C_{H^+} the concentration of hydrogen ions in the OHP (moles per square centimeter), θ the fraction of the surface covered by H_{ad}, ΔG_C^* the chemical part of the free energy of activation, and $\beta(\Delta\phi)F$ the electrical contribution to the free energy of activation, with $\Delta\phi$ is the potential difference between the metal and the OHP. Writing

$$\Delta\phi = \Delta\phi_R + \eta \quad (8.3.8)$$

where $\Delta\phi_R$ is $\Delta\phi$ when the electrode is held at the reversible potential and η is the overpotential, Eq. (8.3.7) becomes

$$v_1 = k'\, \exp(-\Delta G_C^*/RT)\exp(-\beta\eta_1/RT) \quad (8.3.9)$$

where k' is a constant.

Applying the Bronsted law, the new rate v_2 after the adsorption of the organic material and changing the overpotential to a value η_2 will be

$$v_2 = k'\, \exp[-(\Delta G_C^* - \beta\,\Delta(\Delta H))/RT]\exp(-\beta\eta_2 F/RT) \quad (8.3.10)$$

Comparing (8.3.9) and (8.3.10), we have

$$v_2 = v_1\, \exp[\beta\,\Delta(\Delta H)/RT]\exp[-\beta(\Delta\eta)F/RT] \quad (8.3.11)$$

where

$$\begin{aligned}
\Delta\eta &= \eta_2 - \eta_1 \\
&= -0.222 - (-0.160) \\
&= -0.062 \text{ V}
\end{aligned}$$

Taking the symmetry factor β as $1/2$ and substituting numerical values in (8.3.11) yields

$$\begin{aligned}
v_2 &= 2 \times 10^{-7} \exp(-0.5 \times 12{,}000/1.987 \times 298) \\
&\quad \times \exp(0.5 \times 0.062 \times 96{,}500/8.314 \times 298) \\
&= 2.7 \times 10^{-11} \text{ mole cm}^{-2} \text{ sec}^{-1}
\end{aligned}$$

5 The potential energy of a simple harmonic oscillator is parabolic with respect to displacement. The potential energy profile of the charge-transfer reaction is thus appropriately described by two intersecting parabolas as shown in Fig. 8.5.1. The zero-point energy of the initial state oscillator is displaced $-Q$ from the zero-point energy (taken as zero for convenience) of the final state oscillator. The abscissa q represents, in one dimension, the multidimensional displacements of water molecules near the ion.

The reorganizational energy E_s is the energy which must be supplied to the solvated ion in the initial state so that it may have the configuration of the ground state of the product ion. The energy E_s is marked in Fig. 8.5.1. The activation energy E^* is the energy which must be supplied to the reacting ion before electron transfer obeying the Frank–Condon principle can occur, and is shown in Fig. 8.5.1.

Writing U_R and U_P as the energies of the initial- and final-state harmonic oscillators, we have, by inspection of Fig. 8.5.1,

$$U_P = k(q - d)^2 \qquad (8.5.2)$$

$$U_R = kq^2 - Q \qquad (8.5.3)$$

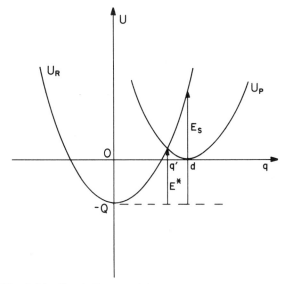

Fig. 8.5.1. Parabolic potential energy curves for initial- and final-state harmonic oscillators.

At the intersection point of the two curves,

$$U_R = U_P \qquad (8.5.4)$$

and

$$q = q' \qquad (8.5.5)$$

Substituting (8.5.4) and (8.5.5) in (8.5.2) and (8.5.3) and subtracting (8.5.3) from (8.5.2) yields

$$kd^2 - 2kq'd + Q = 0 \qquad (8.5.6)$$

and rearranging yields:

$$q' = (Q + kd^2)/2kd \qquad (8.5.7)$$

From Fig. 8.5.1, it is clear that

$$E_s = kd^2 \qquad (8.5.8)$$

and substituting (8.5.8) in (8.5.7) yields

$$k(q')^2 = (Q + E_s)^2/4E_s \qquad (8.5.9)$$

But from Fig. 8.5.1 it is clear that

$$E^* = k(q')^2 \qquad (8.5.10)$$

and substituting (8.5.10) into (8.5.9) yields

$$E^* = (Q + E_s)^2/4E_s \qquad (8.5.1)$$

It is apparent from the above that (8.5.1) does not depend on any particular model of charge transfer since the details of the actual charge-transfer process were neglected in the derivation of (8.5.1). Further, since (8.5.2) and (8.5.3) are the equations to parabolas, it is apparent that (8.5.1) is only valid when the initial and final states of the system approximate harmonic-oscillators.

---

7 The image interaction energy between a univalent ion and a metallic conductor is given by

$$U_{\text{image}} = -e_0^2/4\epsilon r \qquad (8.7.1)$$

where ϵ is the dielectric constant and r is the distance between the metal and the ion situated in the OHP. Assuming the position of the ion can fluctuate by as much as 1 Å, we have for the fluctuation of the image energy

$$\Delta U_{\text{image}} = -(e_0{}^2/4\epsilon \times 10^{-8})(\tfrac{1}{3} - \tfrac{1}{4}) \qquad (8.7.2)$$

taking the metal-to-OHP distance as 4 Å. Substituting numerical values for e_0 and ϵ into (8.7.2), we have

$$
\begin{aligned}
\Delta U_{\text{image}} &= -4.8 \times 10^{-10} \times 300/4 \times 6 \times 10^{-8} \times 12 \\
&= -0.05 \text{ eV} \\
&= -0.05 \times 23.06 \text{ kcal mole}^{-1} \\
&= -1.2 \text{ kcal mole}^{-1}
\end{aligned}
$$

Consequently, the magnitude of the image energy fluctuations is only about 10 % of typical activation energies for charge-transfer processes (10–15 kcal mole^{-1}).

In an elementary charge-transfer reaction in the gas phase, an electron leaves a metal electrode and is captured by an ion situated l cm away from the metal. Calculation of the image energy of the electron as it leaves the metal and calculation of the Coulombic energy of the electron as it approaches the ion provide ways of calculating the energy barrier through which the electron must tunnel to reach the ion.

The image interaction energy between a metal and the electron is given by

$$U_{\text{image}} = -e_0{}^2/4r_M\epsilon \qquad (8.7.3)$$

where r_M is the distance between the metal and the electrode. Since U_{image} becomes very small when r_M is large, the zero of the energy scale is the electron at rest at infinity. Taking the dielectric constant as 1, this interaction energy at a separation of 1 Å is given by

$$
\begin{aligned}
U_{\text{image}} &= 4.8 \times 10^{-10} \times 300/1 \times 10^{-8} \times 4 \\
&= -3.60 \text{ eV}
\end{aligned}
$$

The energy of interaction between an electron and the ion is Coulombic and given by

$$U_{\text{Coul}} = -e_0{}^2/r_I\epsilon \qquad (8.7.4)$$

where r_I is the electron–ion separation. Again the electron at rest at infinity is the zero of the energy scale.

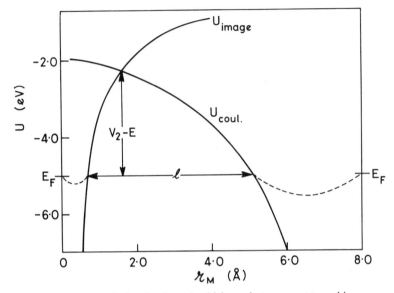

Fig. 8.7.1.   The barrier through which an electron must tunnel in a charge-transfer process.

Figure 8.7.1 shows a plot of the electron energy as a function of the electron–metal separation and also as a function of the metal–ion separation. The distance between the metal and the ion is taken as 8 Å. The energy of the electron in the metal is the Fermi energy, which, on the scale of the electron at rest at infinity, is the negative of the electron work function of the metal. For a nonradiative transfer of an electron from the metal to the ion to take place, there must be vacant electron levels in the ion at the Fermi energy of the metal. The energies of the electron in the metal and of the ion are marked in Fig. 8.7.1.

The intersecting image force curve and Coulombic force curve form a barrier through which the electron must tunnel to the ion. Very close to the metal or the ion, (8.7.3) or (8.7.4) no longer gives a true picture of the energy of the electron since both U_{image} and U_{Coul} become very large and negative when the electron is close. The true situation close to the metal or the ion is better represented by the dotted lines in Fig. 8.7.1, which take into consideration the fact that at the metal or the ion, the electron energy is that of the Fermi energy of the metal.

It is clear from Fig. 8.7.1 that the electron has to tunnel through a barrier of width l equal to 4.4Å and height $V_2 - E$ equal to 2.8 eV. The probability of tunneling P_T through a barrier of width l and height V_2 at a level E is given (roughly) by the Gamow equation as

$$P_T = \exp\{-(4\pi l/h)[2m(V_2 - E)]^{1/2}\} \tag{8.7.5}$$

where m is the mass of the particle. For the units in the exponential to cancel, $V_2 - E$ must be in ergs when m is in grams and l in centimeters.

Substituting numerical values in (8.7.5), we have

$$P_T = \exp\left[- \frac{4\pi \times 4.4 \times 10^{-8}}{6.624 \times 10^{-27}} \right.$$
$$\left. \times (2 \times 9.1 \times 10^{-28} \times 2.8 \times 1.60 \times 10^{-12})^{1/2}\right]$$
$$= 5.6 \times 10^{-4}$$

where the factor of 1.60×10^{-12} converts electron-volts to ergs.

It should be noted that (8.7.5) applies strictly only to a rectangular barrier and for this reason, the above value of P_T is only approximate. A more accurate value of P_T could be obtained by considering an Eckhardt barrier.

| 9 | Consider the homogeneous exchange reaction

$$Fe^{3+} + Fe^{2+} \rightarrow Fe^{2+} + Fe^{3+} \tag{8.9.2}$$

proceeding along the path

$$Fe_{esc}^{3+} + Fe_{esc}^{2+} \rightarrow [Fe_{nesc}^{3+} + Fe_{nesc}^{2+}]$$
$$\downarrow\uparrow \tag{8.9.4}$$
$$Fe_{esc}^{2+} + Fe_{esc}^{3+} \leftarrow [Fe_{nesc}^{2+} + Fe_{nesc}^{3+}]$$

where the subscripts esc and nesc refer to the equilibrium solvent configuration and the nonequilibrium solvent configuration surrounding ions in their normal and activated states respectively. The free energy of activation is the free energy difference between the activated and initial states and is consequently

$$\Delta G_{hom}^* = G_{Fe_{nesc}^{3+}} + G_{Fe_{nesc}^{2+}} - G_{Fe_{esc}^{3+}} - G_{Fe_{esc}^{2+}} \tag{8.9.5}$$

The corresponding heterogeneous reaction

$$Fe^{3+} + e(M) \rightarrow Fe^{2+} \qquad (8.9.3)$$

proceeds along the path

$$Fe^{3+}_{esc} + e(M) \rightarrow [Fe^{3+}_{nesc} + e(M)]$$
$$\downarrow\uparrow \qquad\qquad (8.9.6)$$
$$Fe^{2+}_{esc} \leftarrow [Fe^{2+}_{nesc}]$$

The electrochemical free energies of activation, $\Delta\vec{G}^*_{het}$ and $\Delta\overleftarrow{G}^*_{het}$ for the forward and reverse reactions respectively are given by

$$\Delta\vec{G}^*_{het} = G_{Fe^{3+}_{nesc}} - G_{Fe^{2+}_{esc}} \qquad (8.9.7)$$

and

$$\Delta\overleftarrow{G}^*_{het} = G_{Fe^{2+}_{nesc}} - G_{Fe^{2+}_{esc}} \qquad (8.9.8)$$

Assuming that activation of the homogeneous and heterogeneous processes takes place by the same unspecified mechanism, (8.9.7) and (8.9.8) can be substituted into (8.9.5), leading to the result

$$\Delta G^*_{hom} = \Delta\vec{G}^*_{het} + \Delta\overleftarrow{G}^*_{het} \qquad (8.9.9)$$

However, at the reversible potential, the rates of the forward and reverse heterogeneous reactions are the same, and for equal concentrations of reactants and products,

$$\Delta\vec{G}^*_{het} = \Delta\overleftarrow{G}^*_{het} \qquad (8.9.10)$$

and under these conditions (8.9.9) becomes

$$\Delta G^*_{hom} = 2\Delta G^*_{het} \qquad (8.9.11)$$

The rate constants for the homogeneous and heterogeneous reactions can be written

$$k_{hom} = Z_{hom} \exp(-\Delta G^*_{hom}/RT) \qquad (8.9.12)$$

$$k_{het} = Z_{het} \exp(-\Delta G^*_{het}/RT) \qquad (8.9.13)$$

from which, together with (8.9.11), it follows that

$$(k_{hom}/Z_{hom})^{1/2} = (k_{het}/Z_{het}) \qquad (8.9.1)$$

Since in this derivation of (8.9.1) no assumptions have been made about the mechanism of activation, the experimental verification of (8.9.1) does not constitute evidence in favor of the electrostatic model of electrode processes.

CHAPTER 9

MORE BASIC ELECTRODE KINETICS

[1] (a) Calculate the diffusion limiting current for the oxidation of an organic compound at an electrode in a quiescent solution. Assume six electrons are involved in the reaction. $C_{organic} = 10^{-2}$ mole liter^{-1}; $D_{organic} = 2 \times 10^{-5}$ cm^2 sec^{-1}.

(b) Calculate the transition time for the same system at a current density of 0.30 A cm^{-2}.

[2] For the cathodic reaction

$$A + 2e^- \rightarrow 2B \qquad (9.2.1)$$

$i_0 = 1.10 \times 10^{-3}$ A cm^{-2} and $(\partial\eta/\partial i)_{n\to 0} = -28.3$ ohms cm^2 at 25°C. What is the stoichiometric number of the reaction and what is its mechanistic significance in this case?

[3] (a) Figure 9.3.1 shows galvanostatic transients of current density 35 mA cm^{-2} for a Pt electrode in two solutions. One contains 1.3 μmoles liter^{-1} benzene and the other is a blank. The blank curve corresponds to the formation of an oxide film on the Pt, the other to the oxidation of adsorbed benzene plus the formation of the oxide film. Calculate the charge required to oxidize the adsorbed benzene.

(b) Plots of equilibrium benzene coverage versus potential

121

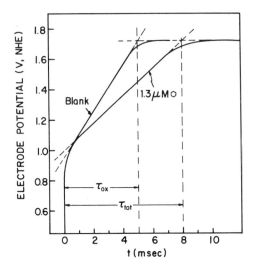

Fig. 9.3.1.   Galvanostatic transients of a platinum
electrode in a 1.3 μmoles liter^{-1} benzene solution and
for a blank solution.

obtained by radiotracer and galvanostatic transient methods are given
in Figs. 9.3.2 and 9.3.3, respectively. Assuming that benzene is adsorbed
intact and that the oxidation takes place according to

$$C_6H_6 + 12H_2O \rightarrow 6CO_2 + 18H^+ + 18e^- \qquad (9.3.1)$$

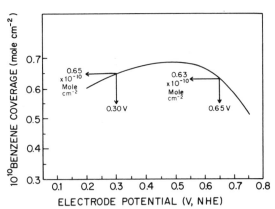

Fig. 9.3.2.   Benzene coverage estimated by radiotracer
techniques on a platinum electrode.

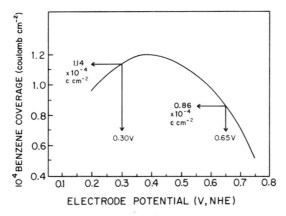

Fig. 9.3.3. Benzene coverage estimated by galvanostatic transient techniques on platinum.

compare numerically the results of these two methods at 0.30 and 0.65 V NHE. What conclusions can be drawn about the principal species present on the surface at the two potentials?

| 4 | (a) The adsorption of propane from 85% H_3PO_4 has been studied by Brummer (1965). At elevated temperatures, the initial adsorption obeys the diffusion equation |

$$Q = 2nF(D/\pi)^{1/2}Ct^{1/2} \qquad (9.4.1)$$

where Q is the anodic charge per square centimeter of geometric area needed in a galvanostatic transient to oxidize the adsorbed material to CO_2 and H^+. The process involves n electrons, and t is the time. If adsorption occurs without partial oxidation, calculate the initial slope of the Q versus $t^{1/2}$ curve using $C_{propane} = 1.6 \times 10^{-4}$ mole liter^{-1} and $D_{propane} = 1.07 \times 10^{-6}$ cm^2 sec^{-1}. What would be the slope per real square centimeter of surface if the roughness factor is three?

(b) The fractional surface area free of adsorbed propane can be measured by cathodic charging with hydrogen:

$$H^+ + e^- \rightarrow H_{ads} \qquad (9.4.1)$$

where each adsorbed hydrogen atom occupies one platinum atom on the surface. At a chosen potential, propane is allowed to adsorb on the electrode for a certain time t and the unoccupied surface is then charged with hydrogen by applying a cathodic pulse. From the number of coulombs passed, the fractional coverage by hydrogen θ_H, which corresponds to the fraction of surface sites not covered by adsorbed propane, is obtained. The experiment is repeated for different times of adsorption of propane. Figure 9.4.1 shows plots of θ_H against $t^{1/2}$ for adsorption of propane at two potentials.

Taking unit coverage by adsorbed hydrogen as 1.3×10^{15} atoms cm^{-2} of platinum, calculate the number of surface sites occupied by one propane molecule at 0.2 V and also at 0.4 V.

<u>5</u> (a) By considering the first step of the hydrogen evolution reaction as a quasi equilibrium

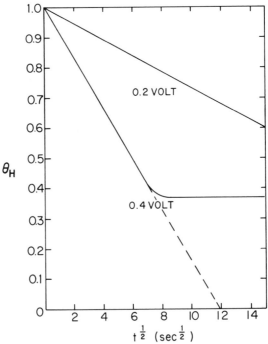

Fig. 9.4.1.   Fractional coverage by hydrogen obtained by cathodic charging experiments as a function of $t^{1/2}$, where t is the time of adsorption of propane in phosphoric acid.

$$H^+ + e^- \rightleftarrows H_{ad} \tag{9.5.1}$$

derive the relation

$$d\theta/dE = -F\theta(1 - \theta)/RT \tag{9.5.2}$$

where θ is the fractional coverage by H_{ad} at potential E.

(b) During galvanostatic transients, the equilibrium (9.5.1) can give rise to large pseudocapacitances C given by

$$C = -F\Gamma \, d\theta/dE \tag{9.5.3}$$

where Γ is the unit coverage by H_{ad} $(1.7 \times 10^{-9}$ atom cm$^{-2})$. The experimental value of the pseudocapacitance on an iron electrode at the potential, -0.45 V NHE, is 230 μF cm^{-2}. Calculate the corresponding fractional coverage by adsorbed hydrogen.

6   Table 9.6.1 lists the standard potentials for the electrode processes that may occur on iron electrodes in aqueous solution. Construct a potential–pH diagram (Pourbaix diagram) for iron. From the diagram, deduce the lower pH limits of stability of iron in (a) oxygen-saturated solution containing 10^{-2} mole liter^{-1} ferric ions, (b) oxygen-free solution containing 10^{-2} mole liter^{-1} ferrous ions. Assume that oxide films provide complete protection when they are thermodynamically stable.

TABLE 9.6.1.   Standard Potentials for the Electrochemical Processes Occurring at Iron

Reaction	Standard potential, V
$Fe^{2+} + 2e^- \rightleftarrows Fe$	-0.44
$Fe^{3+} + 3e^- \rightleftarrows Fe$	-0.04
$Fe_3O_4 + 8H^+ + 2e^- \rightleftarrows 3Fe^{2+} + 4H_2O$	0.98
$Fe_3O_4 + 8H^+ + 8e^- \rightleftarrows 3Fe + 4H_2O$	0.08
$3Fe_2O_3 + 2H^+ + 2e^- \rightleftarrows 2Fe_3O_4 + H_2O$	0.22
$2H^+ + 2e^- \rightleftarrows H_2$	0.00
$O_2 + 4H^+ + 4e^- \rightleftarrows 2H_2O$	1.23
$Fe_2O_3 + 6H^+ + 2e^- \rightleftarrows 2Fe^{2+} + 3H_2O$	0.73
$2Fe^{3+} + 3H_2O \rightleftarrows Fe_2O_3 + 6H^+$	$*$

$*$ $pK_s = 0.72$, where K_s is the solubility product.

> [7] Using the following data, construct an E–log i diagram (Evans–Hoar diagram) for the corrosion of iron in hydrogen-saturated oxygen-free solution of $pH = 3.1$ and $a_{Fe^{2+}} = 0.02$ (molar scale).

For iron dissolution, $i_0 = 9 \times 10^{-7}$ A cm^{-2}; anodic $d\eta/d(\log i) = 0.04$ V; the corrosion potential of iron is -0.215 V (RHE) and $E^\circ_{Fe^{2+}/Fe} = -0.44$ V. For the hydrogen evolution reaction (h.e.r.), the cathodic $d\eta/d(\log i) = -0.12$ V.

Calculate (a) the corrosion rate, (b) the exchange current density for the hydrogen evolution reaction on iron, (c) the cathodic protection current required to reduce the corrosion rate to zero.

> [8] One preparation method for aniline depends on the reduction of nitrobenzene with hydrogen using colloidal platinum as a catalyst. Examine the kinetics of the process assuming that it occurs by an electron-transfer mechanism at the platinum. Neglect the polarization of the hydrogen evolution reaction and take the following parameters for the reduction of nitrobenzene: $i_0 = 10^{-8}$ A cm^{-2}; $d\eta/d(\log i) = -0.12$ V; and the reversible potential is 0.87 V (RHE).

Calculate the rate of reduction in moles per liter per second if there are 10^{10} particles of Pt per liter of average diameter 2×10^{-4} cm.

> [9] Potential sweeps can yield transient data at high sweep rates, and steady-state data at low sweep rates.

(a) Estimate the minimum sweep rate that can be used to measure the coverage of a Pt electrode by benzene at 50°C before readsorption contributes an error of 10 %. Assume adsorption follows the equation

$$\frac{d\theta}{dt^{1/2}} = \frac{2\theta_{eq}}{K\Gamma_{max}} \left(\frac{D}{\pi} \right)^{1/2} \qquad (9.9.1)$$

where K is the equilibrium constant for adsorption and equals 6×10^8 cm^3 mole^{-1}; Γ_{max} is the coverage corresponding to $\theta = 1$ and equals 2.5×10^{-10} mole cm^{-2}; θ_{eq} is the equilibrium coverage and may be taken as 0.3; $D = 1.0 \times 10^{-5}$ cm^2 sec^{-1}.

(b) The oxidation of ethylene in sulfuric acid solution has been investigated by the potential sweep technique. The apparent $d\eta/d(\log i)$ on platinum at 80°C increases from the steady-state value of 0.145 V to about 0.20 V as the sweep rate goes from 10^{-5} V sec^{-1} to 10^{-3} V sec^{-1} and thereafter remains constant. The Tafel region extends over a span of 0.35 V. Show that the higher slope is consistent with an irreversible diffusion-controlled reaction for which (Delahay, 1953)

$$i = (\pi D\lambda)^{1/2} nFC\chi(\lambda t) \tag{9.9.2}$$

where $\lambda = (\alpha F/RT)v$; n is the number of electrons transferred in the overall reaction; α is the transfer coefficient; v is the sweep rate in volts second^{-1}; $\chi(\lambda t)$ is a couple function of (λt) and is plotted in Fig. 9.9.1.

10 For the ethylene oxidation reaction on platinum at 80°C, the following parameters are found over the pH range 0.5–12.5 and the potential range 0–1 V (RHE):

$$\Delta H \approx 20 \text{ kcal mole}^{-1}; \qquad \epsilon_{CO_2} = 100\%; \qquad [\partial E/\partial(\ln i)]_p = 2RT/F$$

$$(\partial i/\partial p)_E < 0; \qquad di_L/dp \approx 1$$

ϵ_{CO_2} is the Coulombic efficiency for oxidation of ethylene to CO_2, p is

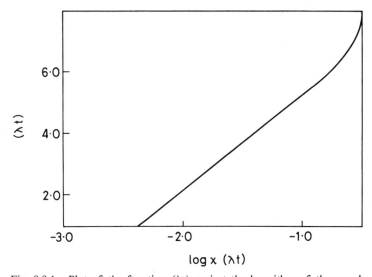

Fig. 9.9.1. Plot of the function (λt) against the logarithm of the couple function $\chi(\lambda t)$. (Gileadi et al., 1966.)

the partial pressure of ethylene, and i_L is the diffusion-limited current. If the same mechanism applies in both acid and alkaline solution, what is the rate-determining step prior to diffusion control? Consider only (OH) and (C_2H_4) as possible adsorbed species.

---

11 For the oxalic acid oxidation reaction on platinum at 80°C, the following empirical relationship was found in acid solution:

$$i = k(C_{ox})^{0.35}(C_{H^+})^{0.55} \exp(FV/RT) \qquad (9.11.1)$$

Assume that the reacting species is the undissociated acid; the only adsorbed species that need to be considered are OH, HC_2O_4, and HCO_2; charge-transfer steps are always single-electron steps. Then write two possible reaction paths for each of the following possibilities: (a) OH radicals are not involved; (b) the oxalic acid molecule undergoes a chemical reaction step:

$$H_2C_2O_4 + OH_{ad} \rightarrow H_2O + (HC_2O_4)_{ad} \qquad (9.11.2)$$

and thereafter OH radicals are not involved; (c) two OH radicals are involved per molecule of oxalic acid. Assuming Temkin adsorption conditions, establish the r.d.s. (if any) for each path that is consistent with the observed potential dependence of (9.11.1), and then show which mechanisms can be eliminated on the basis of the other data.

---

12 The kinetic parameters in Table 9.12.1 were observed for the anodic oxidation of each of four saturated hydrocarbons on platinum in 85 % phosphoric acid at potentials below 0.5 V RHE.

Write down about a dozen partial mechanism hypotheses (sequences up to and including the rate-determining step) and tabulate the values for the three parameters above predicted by each mechanism

TABLE 9.12.1. Kinetic Parameters for the Anodic Oxidation of Saturated Hydrocarbons on Platinum

ϵ_{CO_2} , %	$(\partial E/\partial \ln i)_{p,T}$	$\partial i/\partial a_{H_2O}$	$(\partial \ln i/\partial \ln p)_{a_{H_2O}}$
89–102 $\pm$ 5	RT/F	0	1

assuming Langmuir adsorption conditions and low coverages by adsorbed species. By comparing the parameters with experimental values, reduce the number of possible mechanisms to two. Estimate which of the two would be energetically unfavorable from the following data. Bond strengths are: C–C = 93.1 kcal mole^{-1}; C–H = 106 kcal mole^{-1}; Pt–C = 44 kcal mole^{-1}; Pt–H = 62 kcal mole^{-1}; also, $\Delta H_{ad}(H_2O)$ = 20 kcal mole^{-1}; the ionization energy of H, I_{H-H^+} = 313 kcal mole^{-1}; $\Delta H(H_g^+ \rightarrow H_{soln}^+)$ = -263 kcal mole^{-1}; the work function of platinum Φ_{Pt} = 123 kcal mole^{-1}. Write down the most probable mechanism for the oxidation of saturated hydrocarbons up to the rate-determining step and give one possibility for the rest of the reaction.

ANSWERS

1

(a) Under steady-state conditions, the flux J of the organic molecules arriving at the electrode is given by Fick's first law:

$$J = -D \, dC/dx \quad \text{moles cm}^{-2} \text{ sec}^{-1} \qquad (9.1.1)$$

where the concentration is expressed in moles per cubic centimeter and J and dC/dx are vectors directed toward and away from the electrode, respectively. For an oxidation involving six electrons,

$$J = i/6F \qquad (9.1.2)$$

where i is the current density. Assuming a linear concentration gradient,

$$dC/dx = (C_{electrode} - C_{bulk})/\delta \qquad (9.1.3)$$

where δ is the diffusion layer thickness and may be taken as 0.05 cm in unstirred solution. The gradient is a maximum when $C_{electrode} = 0$ and the maximum, or limiting current, is given from (9.1.1)–(9.1.3) as

$$i_L = 6FDC_{bulk}/\delta \qquad (9.1.4)$$

which becomes, substituting given values,

$$i_L = 6 \times 96{,}500 \times 2 \times 10^{-5} \times 10^{-5}/5 \times 10^{-2}$$
$$= 2.3 \text{ mA cm}^{-2}$$

(b) As shown by Sand (1900), the transition time τ is given by

$$\tau = \pi D(zFC/2i)^2 \qquad (9.1.5)$$

where the concentration C is again expressed in moles cm^{-3}. Substituting the given values, we find

$$\tau = \pi \times 2 \times 10^{-5}[6 \times 96,500 \times 10^{-5}/2 \times 3 \times 10^{-1}]^2$$
$$= 3 \times 10^{-3} \text{ sec}$$

[3] (a) Charge passed during a galvanostatic transient of current density i is $i\tau$ (coulombs cm^{-2}), where τ is the time to reach steady state. Therefore, the charge associated with Pt oxidation is $i\tau_{ox}$ and the charge Q associated with benzene oxidation (see Fig. 9.3.1) is

$$Q = i(\tau_{tot} - \tau_{ox}) \qquad (9.3.2)$$

Therefore,

$$Q = 35 \times 10^{-3}(8.0 \times 10^{-3} - 5.0 \times 10^{-3})$$
$$= 1.05 \times 10^{-4} \text{ coulomb } cm^{-2}$$

(b) From radiotracer measurements, the coverage at 0.30 V, $\Gamma_{0.30}$, is (Fig. 9.3.2)

$$\Gamma_{0.30} = 0.65 \times 10^{-10} \text{ mole } cm^{-2}$$

Assuming that benzene is adsorbed without chemical reaction, the oxidation during a galvanostatic transient will produce 18 electrons per molecule:

$$C_6H_6 + 12H_2O \rightarrow 6CO_2 + 18H^+ + 18e^- \qquad (9.3.3)$$

The charge required to oxidize benzene adsorbed at 0.30 V is

$$Q_{0.3} = 18 \times 96,500 \times 0.65 \times 10^{-10} \text{ coulomb } cm^{-2}$$
$$= 1.13 \times 10^{-4} \text{ coulomb } cm^{-2}$$

This value is in close agreement with the value measured by galvanostatic transients (1.14×10^{-4} coulomb cm^{-2}). The benzene molecule is therefore adsorbed intact at this potential. Similarly, the charge required to oxidize benzene at a coverage of 0.63 mole cm^{-2} would be 0.85×10^{-4} coulomb cm^{-2} measured by galvanostatic transients. If one could attribute the discrepancy to partial oxidation of the benzene during adsorption, say to some species $C_6H_nO_m$, then the number of electrons produced per molecule would be given by the equation

$$C_6H_nO_m + (12-m) H_2O \rightarrow 6CO_2 + (n+12-m) H^+ + (n+12-m)e^-$$

The number of electrons $(n'+12-m)$ required to account for the discrepancy is $(0.86/1.09) \times 18$ or about 14 in place of 18 for benzene.

---

5 (a) For the equilibrium, Eq. (9.5.1), the rate in the forward direction equals that in the reverse direction, and, making use of the Butler–Volmer equation, we have

$$\vec{k}C_{H^+}(1-\theta) \exp(-\beta EF/RT) = \overleftarrow{k}\theta \exp[(1-\beta) EF/RT] \quad (9.5.4)$$

the equilibrium constant is

$$K = \vec{k}/\overleftarrow{k} \quad (9.5.5)$$

and substituting (9.5.4) into (9.5.5) yields

$$K = [\theta/C_{H^+}(1-\theta)] \exp(EF/RT) \quad (9.5.6)$$

Differentiating with respect to E,

$$-(KC_{H^+}/\theta^2) \, d\theta/dE = (F/RT) \exp(EF/RT) \quad (9.5.7)$$

and substituting for K yields (9.5.2):

$$d\theta/dE = -F\theta(1-\theta)/RT$$

(b) Substituting (9.5.3) into (9.5.2) yields

$$C = F^2\Gamma\theta(1-\theta)/RT \quad (9.5.8)$$

and rearranging and substituting numerical values gives

$$\theta(1-\theta) = \frac{230 \times 10^{-6} \times 8.314 \times 298}{(96,500)^2 \times 1.7 \times 10^{-9}}$$
$$\theta = 0.04$$

Standard potentials are on the scale of the standard hydrogen electrode. For this calculation, it is convenient to have all potentials on the reversible hydrogen electrode scale (RHE), i.e., the potential of the electrode measured against a hydrogen electrode in the same solution:

$$E_{Fe^{2+}/Fe}(RHE) = E^{\circ}_{Fe^{2+}/Fe} + (RT/2F)\ln a_{Fe^{2+}} + (RT/F)pH \quad (9.7.1)$$
$$= -0.44 - 0.050 + 0.183$$
$$= -0.307 \text{ V}$$

The Evans–Hoar diagram is shown in Fig. 9.7.1. The iron dissolution line can be drawn since its slope (0.040 V/decade of current density) and one point on the line $[i_0, E_{Fe^{2+}/Fe}(RHE)]$ are known. The corrosion potential is a mixed potential with iron dissolution and hydrogen evolution proceeding at equal and opposite rates. Consequently, the hydrogen evolution line can be drawn with a slope of 0.12 V/decade of current density through the point where the iron dissolution line crosses the corrosion potential, -0.215 V.

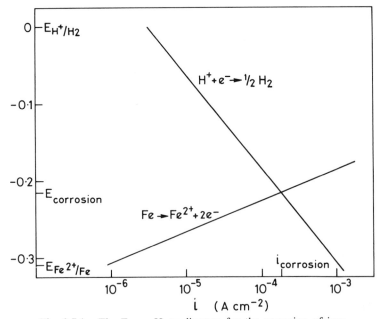

Fig. 9.7.1. The Evans–Hoar diagram for the corrosion of iron.

From Fig. 9.7.1, the corrosion rate of iron in this solution can be seen to be 1.8×10^{-4} A cm^{-2} and i_0 for the h.e.r. is 3.0×10^{-6} A cm^{-2}. The dissolution of the metal ceases when its potential is lowered to its reversible potential, a cathodic current density of 1.2×10^{-3} A cm^{-2}, called the cathodic protection current density, being required to maintain this potential.

---

[9] (a) To cause an error of 10%, the sweep must be of sufficient duration for $\theta_{eq}/10$ to readsorb. A relation between θ and t can be obtained by integrating (9.9.1). The constant of integration is zero because θ is zero at $t = 0$:

$$\theta = (2\theta_{eq}/K\Gamma_{max})(Dt/\pi)^{1/2} \tag{9.9.3}$$

Substituting $\theta_{eq}/10$ in (9.9.3) for θ yields

$$1/10 = (2/K\Gamma_{max})(D/\pi)^{1/2}t^{1/2} \tag{9.9.4}$$

and rearranging gives

$$t = (K\Gamma_{max}/20)^2\pi/D \tag{9.9.5}$$

Substituting given values yields

$$t = (6 \times 10^8 \times 2.5 \times 10^{-10}/20)^2\pi/10^{-5}$$
$$= 17.6 \text{ sec}$$

A reasonable amplitude for a voltage sweep is 1.5 V, thus the minimum sweep rate would be $\sim$0.1 V sec^{-1}.

(b) The Tafel slope $\partial V/\partial(\ln i)$ can be written

$$\frac{\partial V}{\partial(\ln i)} = \frac{\partial V}{\partial[\ln \chi(\lambda t)]}\frac{\partial[\ln \chi(\lambda t)]}{\partial(\ln i)} \tag{9.9.6}$$

But from Eq. (9.9.2), $\partial[\ln \chi(\lambda t)]/\partial(\ln i) = 1$, and (9.9.6) becomes

$$\frac{\partial V}{\partial(\ln i)} = \frac{\partial V}{\partial[\ln \chi(\lambda t)]} \tag{9.9.7}$$

$$= \frac{\partial(\lambda t)}{\partial[\ln \chi(\lambda t)]}\frac{\partial V}{\partial(\lambda t)} \tag{9.9.8}$$

Now, the voltage at any time t during a voltage sweep of rate v is given by

$$V = V_0 + vt \qquad (9.9.9)$$

where V_0 is the initial voltage. From (9.9.8) and the definition of λ, we have

$$\lambda t = (\alpha F/RT)(V - V_0) \qquad (9.9.10)$$

Therefore,

$$\partial V/\partial(\lambda t) = RT/\alpha F \qquad (9.9.11)$$

Now, the steady-state $d\eta/d(\log i)$ of 0.145 V at 80°C is consistent with $\alpha = 0.5$. λt may be estimated from (9.9.10). Substituting numerical values we get

$$\lambda t = 96,500 \times 0.35/2 \times 8.314 \times 353$$
$$= 5.75$$

At $\lambda t = 5.75$, the slope of the plot in Fig. 9.9.1 is 3.1 and hence

$$\partial(\lambda t)/\partial[\ln \chi(\lambda t)] = 3.1/2.303 \qquad (9.9.12)$$

Substituting (9.9.11) and (9.9.12) into (9.9.8) yields

$$\partial V/\partial(\ln i) = 3.1RT/2.303\alpha F$$

and the anodic slope, $\partial V/\partial(\log i)$, is given by

$$\partial V/\partial(\log i) = 3.1 \times 8.314 \times 353/0.5 \times 96,500$$
$$= 0.19 \text{ V}$$

Consequently, the apparent Tafel slope at high sweep rates is consistent with an irreversible diffusion-controlled process.

CHAPTER 10

SOME ELECTRODIC REACTIONS OF INTEREST

1 From the data of Table 10.1.1 for the hydrogen evolution reaction, calculate the relative electrocatalytic activity of an iron electrode with respect to that of a nickel electrode.

TABLE 10.1.1. Kinetic Parameters for Hydrogen Evolution

Electrode	Overpotential at a fixed current density, V	$d\eta/d(\log i)$, V
Fe	−0.427	−0.120
Ni	−0.350	−0.112

2 Galvanostatic transient experiments are performed on the deposition of silver from aqueous solution near the reversible potential. The mechanism is

$$Ag^+ + e^- \rightarrow Ag_{ad} \qquad (10.2.1a)$$

$$Ag_{ad} \rightarrow Ag_{lattice} \qquad (10.2.1b)$$

where step (b) is the diffusion of the adion Ag_{ad} to a growth site where it is rapidly incorporated into the lattice. If step (b) is the rate-determining step (r.d.s.), the overpotential η_D will be given by

$$-\eta_D = \tau_{SD} i RT/F^2 C_0 \qquad (10.2.2)$$

where τ_{SD} is rise time (measured after the double-layer charging is

135

complete) of the galvanostatic transient of current density i; C_0 is the equilibrium concentration of adions and the listed values in Table 10.2.1 are constant within experimental error. Use an average value of C_0 to calculate η_D at each current density. Calculate the overpotential for step (a) rate-determining on the basis of an exchange current density of 0.30 A cm^{-2}. Which step is the r.d.s.?

TABLE 10.2.1.   Equilibrium Adion Concentration C_0 in the Electrodeposition of Silver at Near-Equilibrium Conditions at 25°C

Current density i, mA cm^{-2}	Equilibrium adion concentration C_0, mole cm^{-2}	Rise time τ_{SD}, μsec
0	3.3×10^{-11}	—
6.3	2.1×10^{-11}	187
8.9	2.6×10^{-11}	156
15.0	2.3×10^{-11}	96
22.6	3.2×10^{-11}	75
44.7	5.8×10^{-11}	53

3 Consider the hydrogen evolution reaction

$$H^+ + e^- \rightleftarrows H_{ad} \qquad (10.3.1a)$$

$$H_{ad} + H^+ + e^- \rightarrow H_2 \qquad (10.3.1b)$$

where, on the medium-overvoltage metals, step (b) is the r.d.s., show that, for low θ (fractional coverage by H_{ad}),

$$(d\theta/d\eta) = -(FK/RT)\exp(-F\eta/RT) \qquad (10.3.2)$$

and that at high coverage by H_{ad}, i.e., $\theta \to 1$,

$$d\theta/d\eta \to 0$$

4 The study of certain electrode reactions requires solutions of extreme purity. Estimate the electrode area required to purge 190 ml of solution by anodic preelectrolysis of an impurity initially present at 5×10^{-6} M within 24 hr. Take the maximum acceptable impurity content as 10^{-11} M, the diffusion coefficient of the impurity as 10^{-5} cm^2 sec^{-1}, and the diffusion layer thickness as 0.05 cm. What time would be required if the solution was stirred or circulated over the electrode?

5 Tafel slopes at low overpotentials for the oxygen reaction

$$2H_2O \rightleftarrows O_2 + 4H^+ + 4e^- \qquad (10.5.1)$$

on iridium are $2RT/3F$ (anodic) and $-2RT/F$ (cathodic). At higher anodic overpotentials, the Tafel slope becomes $2RT/F$. (a) What is the stoichiometric number at low overpotential? (b) Show that the low overpotential data are consistent with the "electrochemical oxide" mechanism with the second step as the r.d.s.:

$$H_2O \overset{K}{\rightleftarrows} OH_{ad} + H^+ + e^- \qquad (\theta_{OH} \ll 1) \qquad (10.5.2a)$$

$$OH_{ad} \underset{k_r}{\overset{k_f}{\rightleftarrows}} O_{ad} + H^+ + e^- \qquad (r.d.s.) \qquad (10.5.2b)$$

$$2O_{ad} \overset{K'}{\rightleftarrows} O_2 \qquad (\theta_O \ll 1) \qquad (10.5.2c)$$

(c) Which step as the r.d.s. is consistent with the high anodic overpotential data?

6 Draw two intersecting Morse-type curves for the species M–H and H^+–(H_2O), respectively, to represent the initial and final states of the rate-determining step

$$H^+-(H_2O) + e^- \rightarrow M-H + H_2O$$

where M–H represents a hydrogen atom chemisorbed on the metal electrode and H^+–(H_2O) a hydrated proton in the OHP.

(a) How will the heat of activation vary with the heat of adsorption of hydrogen? If the intersecting parts of the curves can be approximated by straight lines, show geometrically that

$$\tan \gamma/(\tan \lambda + \tan \gamma) = \beta \qquad (10.6.2)$$

where β is the symmetry factor and γ is the angle between the M–H line and the vertical; λ is the angel between the H^+–(H_2O) line and the vertical.

(b) What will the corresponding relationship be if the r.d.s. is the next step?

$$H^+\text{–}(H_2O) + M\text{–}H + e^- \rightarrow H_2 + H_2O + M \qquad (10.6.3)$$

(c) Express the results of (a) and (b) in a qualitative statement about the respective effects of adsorption energy of hydrogen on the metal on the exchange current density of the reaction.

$\boxed{7}$ Table 10.7.1 lists the kinetic parameters of processes that can occur during the deposition of iron from acidified 0.05 M ferrous solution. Assuming activation control, calculate the pH at which the iron deposition current is ten times that of water discharge. Find graphically the potential which gives rise to the maximum efficiency for iron deposition at this pH.

TABLE 10.7.1.   Kinetic Parameters for Iron Deposition in Acidified 0.5 M Fe^{2+} Solution

Reacting species	E°, V NHE	$\log i_0$	$d\eta/d(\log i)$	$\log i_L$
Fe^{2+}	−0.44	$pH - 9$	0.12	−2
H_3O^+	0.0	$-(0.5\,pH + 4)$	0.12	$-pH$
H_2O	0.0	−8.0	0.12	(very large)

$\boxed{8}$ (a) From a consideration of the electrode-kinetic condition for equilibrium in the simple charge-transfer redox reaction

$$A^z + e^- \rightarrow A^{z-1} \qquad (10.8.1)$$

show that the Fermi level in the electrode at the reversible potential is the independent of the electrode material. Note that the Fermi level is equivalent to the electrochemical potential of electrons in the electrode material.

(b) In spite of the fact that the free energy of activation of a simple charge-transfer redox reaction is independent of the electrode material, a roughly linear relationship has been observed between log i_0 and Φ (the electron work function) for the reaction

$$Fe^{3+} + e^- \rightarrow Fe^{2+} \tag{10.8.2}$$

for a series of metals. What sign would you expect for the slope $d(\log i_0)/d\Phi$ if the relationship arose from the effect of the electrokinetic potential upon ionic concentrations in the Helmholtz plane?

9 (a) By integrating the steady-state diffusion equation expressed in spherical coordinates

$$\nabla^2 C = \frac{1}{r^2} \frac{\partial(r^2 \partial C/\partial r)}{\partial r} = 0 \tag{10.9.1}$$

between r_0 and $r_0 + \delta$, where r_0 is the radius of a spherical electrode, δ the diffusion layer thickness, and C the concentration of the diffusing species at r, show that the maximum rate of diffusion in terms of the limiting current i_L is given by

$$i_L/zF = 2DC_\infty/r_0 \tag{10.9.2}$$

where C_∞ is the concentration of the diffusing species in the bulk of the solution.

(b) The overpotential at the tip of a metal dendrite growing by electrodeposition has three components: activation η_A, diffusion η_D, and Kelvin overpotential η_K. The Kelvin contribution arises because the surface energy of the sharp tip alters the reversible potential of the electrode reaction:

$$\eta_K = 2\gamma V/zFr_0 \tag{10.9.3}$$

where γ is the surface energy and V the molar volume of the deposit. For a fixed total overpotential η, derive the following expression for the optimum radius of curvature of the tip of a dendrite for the most rapid growth rate:

$$r_0 = \frac{1 \pm [1 + DC_\infty z^2 F^2 \eta/2\gamma V i_0]^{1/2}}{zF\eta/2\gamma V} \tag{10.9.4}$$

Assume

$$i = i_0\eta_A F/RT \tag{10.9.5}$$

where η_A is the activation overpotential.

(c) Using the data given below for the growth of zinc dendrites, calculate the radius of the tip of the dendrite.

(d) Calculate the ratio of the current density on the dendrite tip to that on the planar part of the electrode.

$$\eta = 0.02 \text{ V}; \qquad D = 7 \times 10^{-6} \text{ cm}^2 \text{ sec}^{-1}$$
$$C_\infty = 10^{-1} \text{ M}; \qquad \gamma = 2000 \text{ ergs cm}^{-2}$$
$$V = 9.2 \text{ ml mole}^{-1}; \qquad T = 35°C$$
$$\delta = 5 \times 10^{-2} \text{ cm}; \qquad i_0 = 0.10 \text{ A cm}^{-2}$$

10 The rate of ethylene oxidation at a given potential on a series of metals and alloys is shown in Fig. 10.10.1. The mechanism of the reaction is

$$C_2H_4 \rightleftarrows (C_2H_4)_{ad} \qquad (10.10.1a)$$

$$H_2O \rightleftarrows OH_{ad} + H^+ + e^- \qquad (10.10.1b)$$

$$(C_2H_4)_{ad} + (OH)_{ad} \rightleftarrows (C_5HtO)_{ad} \qquad (10.10.1c)$$

and the species $(C_2H_5O)_{ad}$ further reacts to produce CO_2 and water. Step (b) is the r.d.s. on platinum, but on all other substrates, it is step (c). The abscissa in Fig. 10.10.1 is the latent heat of sublimation of the metal, a parameter related to the metal–metal bond strength.

Rationalize the volcano shape of the curve in terms of Temkin adsorption of OH radicals and competition between OH radicals and C_2H_4 for adsorption sites.

11 By a consideration of the work of Damjanovic, Dey, Genshaw, Hoare, Huq, Krasilchikov, Rao, Rosenthal, Schumilowa, and Veselovski, discuss:

(a) The dependence of the rate of oxygen reduction upon the degree to which the surface is covered with oxide.

(b) The dependence of exchange current density upon the M–O bond strength.

(c) The likely rate-determining steps for O_2 dissolution on platinum.

What directions appear to be the most promising for fundamental research on the catalysis of oxygen reduction ?

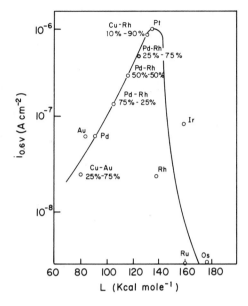

Fig. 10.10.1. Rate of the electrochemical oxidation of ethylene on metal and alloy electrodes as a function of the latent heat of sublimation of the electrode material. (Data from Kuhn *et al.*, 1967.)

12 Investigate the mechanism of the electrochemical oxidation of hydrocarbons, with special reference to the work of Bagotsky, Breiter, Brummer, Grubb, Johnson, Kuhn, Niedrach, Piersma, and Wroblowa.

Give the best available replies to the following questions:

(a) In view of the projected rapid development of atomic power, what is the economic reason for interest and research in the electrochemical oxidation of hydrocarbons? Is there a sociological advantage to be looked for in such research?

(b) Is adsorption of organic materials from solution onto electrodes usually with or without dissociation? How may adsorption with dissociation be detected?

(c) In steady-state measurements, various parameters of the reaction, e.g., stoichiometric number, cathodic and anodic Tafel slopes, and reaction orders, can often be determined and compared with values predicted for various model mechanisms. Transient and potential sweep measurements often lead to information on the nature and surface concentration of intermediates. Discuss the complimentary roles of the two types of measurements for hydrocarbon oxidation.

(d) What generalization seems to be possible with respect to the rate-determining step of organic oxidation mechanisms?

(e) The rational evaluation of nonnoble catalysts for hydrocarbon oxidation depends upon elucidation of the rate-determining steps in hydrocarbon oxidation, and this, in turn, would be greatly helped by knowledge of organic radicals on surfaces. Discuss available and potential approaches to surface radical determination.

(f) What research, then, would you advocate at the present time for furthering our understanding of the mechanism and catalysis of the electrochemical oxidation of hydrocarbons to CO_2?

ANSWERS

$\boxed{1}$ The relative electrocatalytic activity is given by the ratio of the exchange current densities $i_{0(Fe)}/i_{0(Ni)}$. The high-field approximation of the Butler–Volmer equation is, for a cathodic process,

$$i = i_0 \exp(-\beta\eta F/RT) \tag{10.1.1}$$

from which it is apparent that

$$d\eta/d(\log i) = -2.303RT/\beta F \tag{10.1.2}$$

and substituting (10.1.2) into (10.1.1) yields

$$i = i_0 \exp[2.303\eta \, d(\log i)/d\eta] \tag{10.1.3}$$

Since the overpotentials given in Table 10.1.1 refer to a fixed current density, we have, by substituting in (10.1.3) and dividing,

$$\frac{i_{0(Fe)}}{i_{0(Ni)}} = \frac{\exp(2.303 \times 0.350/0.112)}{\exp(2.303 \times 0.427/0.120)}$$
$$= 0.37 \checkmark$$

> **3** With step (10.3.1b) as the r.d.s., then step (10.3.1a) may be considered to be in quasiequilibrium:

$$H^+ + e^- \underset{k_r}{\overset{k_f}{\rightleftharpoons}} H_{ad}$$

and the rates of the forward and reverse reactions are equal. From the Butler–Volmer equation, we have for the equilibrium (10.3.1)

$$k_f C_{H^+}(1 - \theta)\exp(-\beta\eta F/RT) = k_r\theta \exp[(1 - \beta)\,\eta F/RT] \qquad (10.3.2)$$

which rearranges to

$$\theta(1 - \theta)^{-1} = KC_{H^+}\exp(-\eta F/RT) \qquad (10.3.3)$$

where

$$K = k_f/k_r \qquad (10.3.4)$$

Differentiating (10.3.3) with respect to η,

$$[(1 - \theta)^{-1} + \theta(1 - \theta)^{-2}]\,d\theta/d\eta = -(FKC_{H^+}/RT)\exp(-\eta F/RT) \qquad (10.3.5)$$

and rearranging yields

$$[1 + \theta(1 - \theta)^{-1}]\,d\theta/d\eta = -(1 - \theta)(FKC_{H^+}/RT)\exp(-\eta F/RT) \qquad (10.3.6)$$

At low coverage, $(1 - \theta) \to 1$, and (10.3.6) becomes

$$(d\theta/d\eta) = -(FK/RT)\exp(-\eta F/RT) \qquad (10.3.7)$$

At high coverages, $\theta \to 1$, and (10.3.6) becomes

$$d\theta/d\eta = 0$$

> **5** The stoichiometric number ν is given by Parsons (1951) (cf. Bockris and Reddy, 1970) as

$$\nu = \frac{nF}{RT}\left(\frac{\partial \ln i_a}{\partial \eta} - \frac{\partial \ln i_c}{\partial \eta}\right)^{-1} \qquad (10.5.3)$$

where n is the number of electrons transferred in the overall reaction, and $\partial \eta / \partial \ln i_a$ and $\partial \eta / \partial \ln i_c$ are the anodic and cathodic Tafel slopes, respectively. Substituting the given values in (10.5.3) yields

$$\nu = (4F/RT)[(3F/2RT) + (F/2RT)]^{-1}$$

$$= 2$$

(b) For step (10.5.2b) as the r.d.s., the anodic current i_a is given by

$$i_a = 4Fk_f \theta_{OH} \exp[(1 - \beta)\eta F/RT] \qquad (10.5.4)$$

Since step (10.5.2a) may now be regarded as being in quasiequilibrium, we have for this step

$$\theta_{OH}/(1 - \theta_{OH}) = (K/C_{H^+}) \exp(\eta F/RT) \qquad (10.5.5)$$

Since $\theta_{OH} \ll 1$, then $(1 - \theta_{OH}) \approx 1$, and substituting (10.5.5) into (10.5.4) yields

$$i_a = 4F(k_f K/C_{H^+}) \exp[(2 - \beta) \eta F/RT] \qquad (10.5.6)$$

Therefore,

$$\ln i_a + \ln C_{H^+} = \ln 4Fk_f K + (2 - \beta) \eta F/RT \qquad (10.5.7)$$

and differentiating at constant C_{H^+} yields

$$(\partial \eta / \partial \ln i_a)_{C_{H^+}} = 2RT/3F \qquad (10.5.8)$$

which is as observed.

For the cathodic reaction,

$$i_c = 4Fk_r \theta_O C_{H^+} \exp(-\beta \eta F/RT) \qquad (10.5.9)$$

and since step (10.5.2c) may be considered to be in quasiequilibrium:

$$\theta_O^2 /(1 - \theta_O) = K'P_{O_2} \qquad (10.5.10)$$

where P_{O_2} is the partial pressure of oxygen. Hence,

$$i_c = 4Fk_r(K'P_{O_2})^{1/2} C_{H^+}\exp(-\beta \eta F/RT) \qquad (10.5.11)$$

from which it follows that

$$(\partial \eta / \partial \ln i_c)_{P_{O_2}} = -2RT/F \qquad (10.5.12)$$

which is the observed cathodic Tafel slope.

(c) For step (10.5.2a) rate-determining, the anodic current is given by

$$i_a = 4Fk \exp[(1 - \beta) \eta F/RT] \tag{10.5.13}$$

from which it follows that

$$(\partial \eta / \partial \ln i_a)_{C_{H_2O}} = 2RT/F$$

which is consistent with the high-overpotential data.

|7| From the Butler–Volmer equation, we have that the current for iron deposition is given by

$$\log i = \log i_0 - [(V - V_R)/0.12] \tag{10.7.1}$$

where V is the electrode potential and V_R is the reversible potential for iron deposition, on the NHE scale, and

$$V_R = E^\circ + (RT/2F) \ln C_{Fe^{2+}} \tag{10.7.2}$$

since the activity of metallic iron is unity. Substituting numerical values in (10.7.2) yields

$$V_R = -0.44 + (8.314 \times 298/2 \times 96{,}500) \ln(0.05)$$
$$= -0.48 \text{ V}$$

Substituting for V_R and i_0 in (10.7.1) yields

$$\log i_{Fe^{2+}} = p\text{H} - 9.0 - [(V + 0.48)/0.120] \tag{10.7.3}$$

Similarly, for water discharge, we have

$$\log i_{H_2O} = -8.0 - [(V + 0.060p\text{H})/0.120] \tag{10.7.4}$$

since for water discharge

$$V_R = 0.0 + (RT/2F) \ln(C_{H^+})^2 \tag{10.7.5}$$

When

$$i_{Fe^{2+}} = 10i_{H_2O} \tag{10.7.6}$$

and from (10.7.3), (10.7.4), and (10.7.6), we have

$$pH - 9.0 - \frac{V + 0.48}{0.120} = 1.0 - 8.0 - \frac{V + 0.060pH}{0.12} \quad (10.7.7)$$

and condition (10.7.6) is satisfied, independent of potential, when $pH = 4.0$.

Under mixed activation and diffusion control, the current density for iron deposition is related to the potential by

$$V = V_R - \frac{d\eta}{d(\log i_{Fe^{2+}})} [\log i_{Fe^{2+}} + (9.0 - pH)]$$

$$+ \frac{RT}{2F} \ln \left(1 - \frac{i_{Fe^{2+}}}{i_{L-Fe^{2+}}}\right) \quad (10.7.8)$$

and substituting numerical values into (10.7.8) yields

$$V = -0.48 - 0.120(\log i_{Fe^{2+}} + 5.0) + 0.030 \log[1 - (i_{Fe^{2+}}/10^{-2})] \quad (10.7.9)$$

and the potential corresponding to various current densities for iron deposition can be found by substituting values for $i_{Fe^{2+}}$ into (10.7.9). The results are plotted in Fig. 10.7.1.

At a potential of 0.50 V NHE, the overpotential for discharge of H_3O^+ is given by

$$\eta = V - V_R$$
$$= -0.50 - 0.060pH$$
$$= -0.260 \text{ V}$$

and substituting for η in the Butler–Volmer equation, we have

$$\log i_{H_3O^+} = -(0.5pH + 4.0) + (0.260/0.120)$$
$$= -3.80$$

However, since the diffusion-limited current for H_3O^+ discharge is 10^{-4} A cm^{-2} (Table 10.7.1), discharge of hydroxonium ions will proceed at a current density of 10^{-4} A cm^{-2} at all potentials more negative than -0.50 V. This is shown in Fig. 10.7.1.

Water discharge obeys Eq. (10.7.4) and its potential–current density relation is also plotted in Fig. 10.7.1.

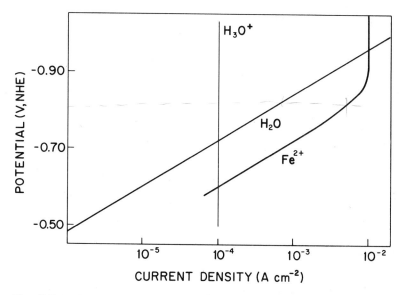

Fig. 10.7.1. Current densities for iron deposition, water discharge, and hydroxonium ion discharge as a function of electrode potential.

The current density efficiency for iron deposition is given by

$$\text{efficiency} = i_{Fe^{2+}}/(i_{Fe^{2+}} + i_{H_2O} + i_{H_3O^+}) \qquad (10.7.10)$$

Interpolating the values of $i_{Fe^{2+}}$, i_{H_2O}, and $i_{H_3O^+}$ from Fig. 10.7.1, the curve of efficiency against potential (Fig. 10.7.2) can be constructed. Under the stated conditions, the maximum efficiency is seen to be 88 % at -0.80 V NHE.

☐9 (a) A general integration of (10.9.1) yields

$$r^2 \, dC/dr = k \qquad (10.9.6)$$

where k is a constant and integration of (10.9.6) between limits r_0 and $r_i = r_0 + \delta$ yields

$$C_{r_i} - C_{r_0} = \int_{r_0}^{r_i} kr^{-2} \, dr \qquad (10.9.7)$$

$$= k[(1/r_0) - (1/r_i)] \qquad (10.9.8)$$

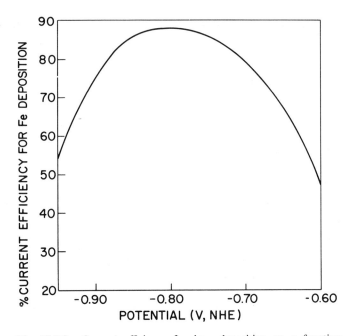

Fig. 10.7.2.   Current efficiency for iron deposition as a function
of electrode potential.

The flux at the electrode J_{r_0} is given by Fick's first law as

$$J_{r_0} = -D(dC/dr)_{(r=r_0)} \qquad (10.9.9)$$

and substituting (10.9.6) and (10.9.8) into (10.9.9) gives

$$J_{r_0} = \frac{D}{r_0^2} \frac{C_{r_i} - C_{r_0}}{(1/r_0) - (1/r_i)} \qquad (10.9.10)$$

Now

$$i = zFJ_{r_0} \qquad (10.9.11)$$

and since the maximum flux across the diffusion layer is obtained when C_{r_0} is zero and $C_{r_i} = C_\infty$ the diffusion limiting current i_L is given by (10.9.2)

$$i_L = zFDC_\infty/r_0$$

(b) The total overpotential η is given by

$$\eta = \eta_A + \eta_D + \eta_K \qquad (10.9.12)$$

where η_A is the activational overpotential given by the Tafel equation and the diffusional overpotential is given by

$$\eta_D = iRT/i_L zF \qquad (10.9.13)$$

Substituting (10.9.3) and (10.9.13) into (10.9.12) yields

$$\eta = \frac{iRT}{i_0 zF} + \frac{iRT}{i_L zF} + \frac{2\gamma V}{zFr_0} \qquad (10.9.14)$$

and differentiating with respect to r_0 yields

$$0 = \frac{di}{dr_0}\frac{RT}{i_0 zF} + \frac{di}{dr_0}\frac{RT}{i_L zF} + \frac{iRT}{zF}\frac{1}{zFDC_\infty} - \frac{2\gamma V}{zFr_0^2} \qquad (10.9.15)$$

The current density will be a maximum when

$$di/dr_0 = 0 \qquad (10.9.16)$$

and substituting (10.9.16) and (10.9.2) into (10.9.15) yields

$$iRT/z^2F^2DC_\infty = 2\gamma V/zFr_0^2 \qquad (10.9.17)$$

Substituting (10.9.17) and (10.9.2) into (10.9.14) yields

$$\eta r_0^2 - \frac{2\gamma Vr_0}{zF} - \frac{2\gamma VDC_\infty}{i_0} - \frac{2\gamma Vr_0}{zF} = 0 \qquad (10.9.18)$$

which is quadratic in r_0 and has the solution (10.9.4):

$$r_0 = \frac{1 \pm [1 + DC_\infty z^2 F^2 \eta/2\gamma Vi_0]^{1/2}}{zF\eta/2\gamma V}$$

(c) By balancing the units in (10.9.4), it can be seen that C_∞ is in moles per milliliter and γ in joules per square centimeter and V in milliliters per mole. Substituting the given values for zinc yields

$$r_0 = 3.7 \times 10^{-5} \text{ cm}$$

(d) Substituting the value of r_0 and other given values into (10.9.17) gives the current density at the tip of the dendrite:

$$i_{\text{tip}} = 2.1 \times 10^{-1} \text{ A cm}^{-2}$$

This value is seen to be considerably less than the diffusion limiting current obtained by substituting numerical values into (10.9.2), i.e., $i_L = 2.7$ A cm^{-2}. Consequently, the growth of dendrites under the present conditions is only partly diffusion-limited.

On the planar part of the surface, the diffusion-limited current i_L^* is given by

$$i_L^* = zFDC_\infty/\delta \tag{10.9.19}$$

and substituting numerical values in (10.9.19) yields

$$i_L^* = 2.7 \times 10^{-3} \text{ A cm}^{-2}$$

If the current on the planar part of the surface was activation controlled, it would be given by (10.9.5),

$$i = (i_0 F/RT) \, | \, \eta_A \, |$$

which yields, on substituting numerical values,

$$i = 7.8 \times 10^{-2} \text{ A cm}^{-2}$$

Consequently, on the planar part of the surface, the current is diffusion controlled with a value of 2.7×10^{-3} A cm^{-2}, and

$$i_\text{tip}/i_\text{plane} = 2.1 \times 10^{-1}/2.7 \times 10^{-3}$$
$$= 78$$

It should be noted that when the current on the dendrite tip is purely diffusion controlled, we have, from (10.9.2) and (10.9.19),

$$i_\text{tip}/i_\text{plane} = \delta/r_0 \tag{10.9.20}$$

and this ratio can become very large. However, r_0 is no longer given by (10.9.4), since in the derivation of (10.9.4), it was assumed that

$$\eta_D = iRT/i_L zF \tag{10.9.21}$$

and (10.9.21) does not hold when $i = i_L$.

CHAPTER 11

SOME ASPECTS OF ELECTROCHEMICAL TECHNOLOGY

1 What principle of thermodynamics determines whether or not a material is stable in contact with a solution? Write explicit equations to show whether or not a metal in contact with a solution is thermodynamically stable if (a) hydrogen ions are available to participate in a cathodic reaction; (b) oxygen is available to participate in a cathodic reaction. According to these criteria, determine whether iron is stable in aqueous solution of $pH = 3$, and whether tin is stable in aqueous solution of $pH = 7$ in contact with air. What essential consideration has been omitted from this treatment of metal stability?

$$E^\circ_{Fe^{2+}/Fe} = -0.44 \text{ V}; \quad E^\circ_{Sn^{2+}/Sn} = -0.136 \text{ V}; \quad E^\circ_{H^+,O_2/H_2O} = 1.23 \text{ V}.$$

2 Answer the following elementary questions concerning electrochemical energy producers:

(a) What is the sign of the free energy change for overall chemical reaction in a working fuel cell?

(b) Write an equation which represents the ideal, maximum efficiency of a fuel cell in terms of the heat of reaction and the cell potential.

(c) In which range do practical fuel cell efficiencies lie? In what range do the efficiencies of internal combustion engines lie?

(d) What is the main *negative* feature of the fuel cell–electric motor combination as a source of mechanical power?

151

| 3 | The corrosion potential of iron in a certain solution is -0.21 V RHE. Assuming that hydrogen enters microvoids in the metal which are approximately spherical and of diameter 10^{-5} cm, determine whether or not the stated conditions will lead the embrittlement of the iron. Take Young's modulus of iron as 1.2×10^{12} dynes cm^{-2} and its surface energy as 1000 ergs cm^{-2}.

| 4 | The emission of unsaturated hydrocarbons from automobile exhausts has been recognized as the principal cause of smog since the later 1960's.

(a) What oil- (or gasoline-) burning systems are alternatives to the internal combustion engine as power sources for automobiles?

(b) Apart from the low level of research on batteries, what is the principal obstacle to the development of a battery-powered automobile?

(c) What other electrochemical power sources can be considered possible replacements for the automobile internal combustion engine?

(d) What is the greenhouse effect? What long-term environmental damage is expected to result from the emissions of the automobile internal combustion engine and other fossil-fuel-consuming power sources?

| 5 | Increasing energy consumption, limited reserves of fossil fuels, and the greenhouse effect will result in the increasing use of nuclear energy. Nuclear generators are only efficient when sufficiently large and heat dissipation necessitates that they be situated on the shore or on platforms in the ocean. Thus, the site of power consumption will frequently be several hundred miles from the nuclear generator.

In a concept known as the "hydrogen economy," the electricity will be used to electrolyse water and the hydrogen produced then pushed through pipes to homes and plants where, in some applications, it may be burned directly to produce heat, and in others converted to electricity at on-site fuel cells.

Use the following information to calculate the cost of producing electrolytic hydrogen in cents/1000 SCF (SCF = 1 ft^3 at 1 atm and 60°F). Neglect in your calculation revenue from sale of oxygen

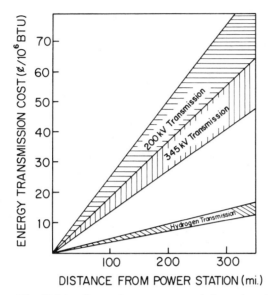

Fig. 11.5.1. Cost of energy transmission plotted against distance from the power station for transmission as electricity and as hydrogen gas.

produced and sale of heavy water produced.* The information below is based on an Allis-Chalmers basic electrolysis cell producing 40×10^6 SCF of hydrogen per day (Gregory et al., 1971).

Cell voltage	1.78 V at 800 A/ft² at 250°F
Electrical energy	0.25¢/kWhr at reactor
Maintenance and operation	4.2¢/1000 SCF hydrogen
Labor and overhead on labor	0.8¢/1000 SCF hydrogen
Desalinated feed water	0.4¢/1000 SCF hydrogen
Depreciation, local tax, and Insurance	9% on capital investment of $11,800,000

Use Fig. 11.5.1 to calculate the cost of supplying heat energy to a consumer 300 miles from the nuclear generator in the form of hydrogen and in the form of electricity transmitted at 345 kV. The

* Heavy water production is estimated at 0.004 lb/1000 SCF of hydrogen. The 1972 market value of heavy water is $28/lb. The market value of both oxygen and heavy water could be expected to fall considerably.

calorific value of hydrogen is 325 Btu/SCF. Local distribution costs are $1.94/10⁶ Btu for electricity and 17¢/10⁶ Btu for natural gas. For hydrogen, the local distribution cost will be somewhat higher, but comparable to that of natural gas.

Compare the cost to this consumer of electricity produced by an on-site hydrogen–air fuel cell of typical efficiency to that directly transmitted. Add 25 % to the cost of the fuel-cell-produced power to cover the amortization of equipment.

How much fresh water per day would be produced by a hydrogen–air fuel cell providing a household with 50 kWhr per day?

6 From a survey of recent publications, compare and contrast the pros and cons of the following electrochemical energy storage devices from the viewpoint of application to automobiles:

(a) Hydrogen–air fuel cell.

(b) Alkali metal–halogen battery.

(c) Alkali metal–sulfur battery.

(d) Zinc–air battery.

(e) Nickel–zinc battery.

(f) Al–Cu cell using propylene carbonate.

7 Consider the feasibility of running a commuter automobile on solar energy. Assume that, at the earth's surface, the power density of solar radiation is about 50 mW cm⁻² over a period of about 12 h. Estimate the amount of energy per day you could expect to collect and store in batteries from clean solar cells covering the roof, trunk, and hood of a compact automobile. The efficiency of good photocells is 10–12 %.

Experience with compact electric vehicles weighing about 2000 lb shows that commuter vehicles of moderate speed, less than 40 mph, require about 1 kWhr per five miles. Assuming that the solar cell output can be electrochemically stored with 80 % efficiency, determine the feasibility of commuting in such a vehicle. Take 30 miles as the average daily distance traveled by commuter automobiles.

8 The direct electrochemical oxidation of cyanide in alkaline solution proceeds according to

$$CN^- \rightarrow CN + e^- \qquad (11.8.1a)$$

$$2CN \rightarrow C_2N_2 \qquad (11.8.1b)$$

$$2OH^- + C_2N_2 \rightarrow CN^- + CNO^- + H_2O \qquad (11.8.1c)$$

where the reaction of cyanogen with hydroxide ions is a homogeneous process taking place in the bulk of the solution. It is desired to use this process to reduce a 50,000 ppm concentration of cyanide in a factory effluent to 0.5 ppm. From free-energy calculations and other considerations, determine whether it would be practical to carry out the oxidation of cyanide in the fuel cell mode or whether the cell would have to be driven. Take the standard free energies of formation of CN^-, C_2N_2, and CNO^- in aqueous solution as 40, 70, and -24 kcal mole^{-1}, respectively.

Sketch out a fluidized-bed type of electrode system appropriate for the purification of the effluent, and discuss the optimization of the operating conditions; current, flow rate, etc. If you conclude that the cell has to be driven, calculate the cost of electricity for processing 1000 gallons of effluent per day.

9 Consider a situation in which a country of 100×10^6 people derives its entire supply of electric power by burning coal of sulfur content 1 %. The electric power consumption is 10 kWhr per person per day.

What is the efficiency of a modern coal-burning power plant and how many tons of coal, of calorific value 13,000 Btu/lb, would be required per year? How many tons of SO_2 would be produced per year and how many tons of H_2SO_4 could be produced from it? Compare this yearly per capita production of sulfuric acid with that of a highly industrialized nation. (The yearly per capita production of sulfuric acid in the United States is about 300 lb.)

Use the data below to decide whether or not it is possible to carry out the SO_2 to H_2SO_4 conversion in the fuel cell mode. Consider the relative merits of obtaining H_2SO_4 in this way compared to those of producing H_2SO_4 from elemental sulfur, the stress being placed on the environmental and economic aspects:

$$\Delta G_f°(H_2SO_4)(aq.) = -177 \text{ kcal mole}^{-1}$$
$$\Delta G_f°(H_2SO_3)(aq.) = -129 \text{ kcal mole}^{-1}$$

|10| In one proposal for dealing with junked cars, the automobile body is shredded in a hammer mill and the scrap compressed into disc-shaped billets 30 cm in diameter and 5 cm thick. The billets are dropped into guides in an electrochemical cell which locate them a few centimeters from stainless steel cathodes. The metals are recovered by anodically dissolving the billet and depositing the metals at the cathodes. Between the anodes and the main cathodes on which iron is deposited are a series of grids held at intermediate potentials to collect the heavy metal components of the junked car. The grids are designed so that they can be replaced periodically and the heavy metals collected. Zinc would be plated out beyond the main cathodes on an electrode held at a more negative potential. Iron would be collected in the form of powder beneath the main cathodes.

Evaluate the technical feasibility of the proposal from the following considerations:

(a) What electrolyte solution would be the most suitable?

(b) What determines the maximum current density for dissolution of the billet? Estimate the numerical value of this current density.

(c) At what potentials should the various cathodes be maintained in order to effect the recovery of all the metallic components of the billet, except aluminum (see Table 11.10.1)?

(d) How can a build-up of ferrous ions in solution due to the codeposition of hydrogen on some of the cathodes be avoided without promoting the reaction

$$Fe^{2+} \rightarrow Fe^{3+} + e^- \tag{11.10.1}$$

on auxiliary anodes?

(e) Make rough estimates of the amount of electricity required to process one car.

(f) How long would it take, at the maximum current density, to process one car? Estimate roughly the size of the cell needed to process 1.6×10^4 junked cars per year.

**TABLE 11.10.1.   Breakdown of Metallic
Components of a Typical Car**

Metal	Amount, lb
Iron	3000
Zinc	54
Aluminum	51
Copper	32
Lead	20
Nickel	5
Chromium	5

11 Discuss factors which would affect your choice of direct energy conversion systems for the following situations:

(a) Powering a torpedo.

(b) Supplying a village in a primitive community with electric power (1 kWhr per person per day).

(c) Powering a car.

(d) Powering a truck.

(e) Powering a large boat.

(f) Providing auxiliary power in space vehicles.

(g) Powering a city.

In each case, give:

(a) A qualitative discussion.

(b) Quantitative calculations in support of your choice taking into consideration economy, power efficiency, and power per unit weight. For example, economy is not important for (a) or (f), but is important for all the others.

(c) A discussion of the effects on the environment, including the greenhouse effect, in the years after 2000 A.D.

12 A relation exists between the cost of power produced at an atomic reactor power station and its size. Investigate this relation

in the literature. How big would the average plant have to be to provide electricity at half the present cost? Suppose the entire supply of U. S. electricity were atomic, at one-fifth the present cost in present dollars, what would be the number of plants which would have to be built and approximately what would the total cost of construction be in present dollars? How long would it take to amortize this investment if the electricity cost is reduced to one-fifth its present cost?

If there were, in this way, a large excess of cheap electricity, to what profitable use could it be put? Consider environmental improvements that could be brought about by that use.

ANSWERS

1 When a metal M is placed in contact with a solution, it will be unstable so long as the process

$$M \rightarrow M^{z+} + ze^- \qquad (11.1.1)$$

where the electrons are retained by the metal, proceeds spontaneously. In the absence of species in the solution capable of accepting the electrons from the metal, (11.1.1) will soon cease as the potential difference built up between the metal and the solution opposes the further transfer of metal ions into the solution. However, if species are available in the solution to accept the electrons, for example, if one of the processes

$$H^+ + e^- \rightarrow \tfrac{1}{2}H_2 \qquad (11.1.2)$$

$$H^+ + \tfrac{1}{4}O_2 + e^- \rightarrow \tfrac{1}{2}H_2O \qquad (11.1.3)$$

or some general cathodic reaction

$$B + e^- \rightarrow B^- \qquad (11.1.4)$$

can occur, it will depolarize the metal and the reaction (11.1.1) can continue to take place. Reactions (11.1.2) and (11.1.3) are the usual depolarizing reactions. Hence, the metal will be unstable if one of the overall processes

$$z\mathrm{H^+} + \mathrm{M} \rightarrow \mathrm{M^{z+}} + \tfrac{1}{2}z\mathrm{H_2} \qquad (11.1.5)$$

$$z\mathrm{H^+} + \tfrac{1}{4}z\mathrm{O_2} + \mathrm{M} \rightarrow \mathrm{M^{z+}} + \tfrac{1}{2}z\mathrm{H_2O} \qquad (11.1.6)$$

proceeds spontaneously, i.e., if it proceeds with a negative change of free energy.

Since reactions (11.1.5) and (11.1.6) are the overall cell reactions for the cells

$$\mathrm{M \mid M^{z+}(aq.) \mid H^+(aq.) \mid H_2} \qquad (11.1.7)$$

$$\mathrm{M \mid M^{z+}(aq.) \mid H^+(aq.)O_2(g) \mid H_2O} \qquad (11.1.8)$$

the free-energy changes can be computed from the cell potentials as

$$-\frac{\Delta G}{zF} = \left[E^\circ_{\mathrm{H^+/H_2}} + \frac{RT}{2F} \ln \frac{(C_{\mathrm{H^+}})^2}{p_{\mathrm{H_2}}} \right] - \left[E^\circ_{\mathrm{M^{z+}/M}} + \frac{RT}{zF} \ln C_{\mathrm{M^{z+}}} \right]$$

$$(11.1.9)$$

$$-\frac{\Delta G}{zF} = \left[E^\circ_{\mathrm{H^+,O_2/H_2O}} + \frac{RT}{4F} \ln(C_{\mathrm{H^+}})^4 p_{\mathrm{O_2}} \right] - \left[E^\circ_{\mathrm{M^{z+}/M}} + \frac{RT}{zF} \ln C_{\mathrm{M^{z+}}} \right]$$

$$(11.1.10)$$

where the activity coefficients have been taken as one.

Consider, now, iron in contact with a solution of $pH = 3$. Initially, the concentration of ferrous ions in the solution will be extremely small and the second term in brackets on the right-hand side of (11.1.9) will be large and negative. Consequently, ΔG will be large and negative and iron dissolution will proceed spontaneously until a finite concentration of ferrous ions is built up in the solution near the electrode. In practice, this concentration can be taken as $10^{-6}\ M$. By a similar argument, the partial pressure of hydrogen $p_{\mathrm{H_2}}$, can be taken as 10^{-6} atm. Substituting numerical values into (11.1.9) yields

$$-\Delta G/zF = [0.0 + 0.0] - [-0.44 - 0.18]$$

$$= 0.62\ \mathrm{V}$$

Consequently, ΔG is negative and the process proceeds spontaneously. Iron is thermodynamically unstable in this solution.

Substituting the numerical values for tin in a solution of $pH = 7$ into (11.1.9) yields

$$-\Delta G/zF = [0.0 - 0.240] - [-0.136 - 0.18]$$
$$= +0.076 \text{ V}$$

Consequently, tin is thermodynamically unstable in this solution due to hydrogen depolarization, although the driving force, -3.5 kcal, is very small. Substituting numerical values into (11.1.10) and taking p_{O_2} as 0.20 yields

$$-\Delta G/zF = [1.23 - 0.43] - [-0.136 - 0.18]$$
$$= 1.11 \text{ V}$$

This shows that tin is thermodynamically unstable in this solution due to oxygen depolarization, (11.1.6), as well.

The essential consideration neglected in this treatment is the consideration of the rates of the processes. Thus, tin is found to not corrode perceptively in solution of $pH = 7$ in contact with air; the depolarization reactions (11.1.5) and (11.1.6) are extremely slow on tin.

| 3 |

The corrosion potential of iron is a mixed potential composed of iron dissolution,

$$\text{Fe} \rightarrow \text{Fe}^{2+} + 2e^- \qquad (11.3.1)$$

and hydrogen evolution,

$$\text{H}^+ + e^- \rightarrow \tfrac{1}{2}\text{H}_2 \qquad (11.3.2)$$

each proceeding at the corrosion potential. For hydrogen evolution, the corrosion potential on the RHE scale, E_{cor} is the overpotential η_H of the process,

$$E_{cor} = \eta_H \qquad (11.3.3)$$

The pressure at which hydrogen is formed *on the surface* of the electrode may be conceived to be that which thermodynamically causes a decrease in the reversible potential by the amount η_H. This pressure can be calculated from the Nernst equation:

$$\eta_H = -(RT/2F) \ln p_{H_2} \qquad (11.3.4)$$

assuming that the h.e.r. takes place by means of proton discharge followed by combination of H atoms.

Rearranging (11.3.4), we have

$$p_{H_2} = \exp(-2\eta_H F/RT) \tag{11.3.5}$$

and substituting numerical values yields

$$p_{H_2} = \exp(2 \times 0.21 \times 96{,}500/8.314 \times 298)$$
$$= 10^7 \text{ atm}$$

Equilibrium is established when the fugacity of hydrogen on the surface of the metal and the fugacity of hydrogen in microvoids inside the metal become equal.

It has been shown that cracks will develop at a microvoid if the pressure inside the microvoid exceeds a certain critical pressure P_{crit} (Beck *et al.*, 1966). Further,

$$P_{crit} = (16\gamma Y/3d)^{1/2} \tag{11.3.6}$$

where γ is the surface energy of the metal, Y is Young's modulus, and d is the diameter of the microvoid. Substituting the given values for iron, we have

$$P_{crit} = (16 \times 10^3 \times 1.2 \times 10^{12}/3 \times 10^{-5})^{1/2}$$
$$= 2.5 \times 10^{10} \text{ dynes cm}^{-2}$$
$$= 2.5 \times 10^4 \text{ atm}$$

Consequently, when the fugacities of the hydrogen on the surface and in the microvoids are equal, the pressure of hydrogen in the microvoids would be considerably in excess of P_{crit} and the stated conditions will lead to the propagation of cracks and the embrittlement of the iron.

[5] To determine the cost of hydrogen production, it is first necessary to calculate the number of moles of hydrogen in 1 SCF and the amount of electrical power required to produce 1 SCF of hydrogen:

$$1 \text{ SCF} = 273 \times 28.3/288.4 \quad \text{liters (NTP)}$$
$$= 273 \times 28.3/288.4 \times 22.4 \quad \text{moles}$$

The evolution of 1 SCF of hydrogen therefore requires:

$$273 \times 28.3 \times 2 \times 96{,}500/288.4 \times 22.4 = 23.2 \times 10^4 \quad \text{coulombs}$$

Since the cell voltage is 1.78 V, the electrical energy required is

$$23.2 \times 10^4 \times 1.78/1000 \times 3600 = 0.115 \text{ kWhr/SCF}$$

and the cost of electrical power to produce 1000 SCF of hydrogen is

$$0.115 \times 10^3 \times 0.25 = 28.7 \text{¢}/1000 \text{ SCF}$$

The total cost of producing hydrogen can be arrived at by summing the cost of electrical power, the cost of services, and the depreciation. The latter, per 1000 SCF of hydrogen, is given (see Table 11.10.1) by

$$11.8 \times 10^6 \times 100 \times 0.09/40 \times 10^3 \times 365 = 7.3 \text{¢}/1000 \text{ SCF}$$

Therefore, the total cost of hydrogen production is

$$28.7 + 7.3 + 4.2 + 0.8 + 0.4 = 41.4 \text{¢}/1000 \text{ SCF}$$
$$= 41.4/0.325 \text{¢}/10^6 \text{ Btu}$$
$$= 127 \text{¢}/10^6 \text{ Btu}$$

The total cost of supplying heat energy to a consumer in the form of hydrogen is the sum of the production, transmission (see Fig. 11.5.1), and local distribution costs. Taking the latter at 20¢/10^6 Btu, we have that the total cost is

$$127 + 15 + 20 = 162 \text{¢}/10^6 \text{ Btu}$$

For heat energy in the form of electricity, we have that 10^6 Btu require

$$0.293 \times 10^6/1000 \quad \text{kWhr}$$

and are produced at a cost, at the reactor, of

$$0.293 \times 10^6 \times 0.25/1000 = 73.3 \text{¢}/10^6 \text{ Btu}$$

The total cost of heat energy to a consumer 300 miles from the reactor is obtained by summing the cost of electrical power at the reactor, cost of transmission at 345 kV (Fig. 11.5.1), and the cost of local distribution:

$$73.3 + 57 + 194 = 324 \text{¢}/10^6 \text{ Btu}$$

Consequently, for the supply of heat energy, hydrogen enjoys a 2:1 cost advantage over directly transmitted electrical energy.

For the supply of electrical energy, the cost of directly transmitted electric power will be the same at 324¢/10^6 Btu. The cost of producing electrical energy by burning hydrogen in a hydrogen–air fuel cell will include the cost of hydrogen at the site, the efficiency of the fuel cell, and the depreciation of the fuel cell. Since hydrogen–air fuel cells are typically 75 % efficient, we have the cost of electrical energy per 10^6 Btu as

$$162 \times 1.25/0.75 = 270¢/10^6 \text{ Btu}$$

where the factor of 1.25 covers the amortization of the fuel cell.

Thus, electrical energy produced from hydrogen enjoys a 1.2:1 cost advantage over directly transmitted electrical energy for a consumer 300 miles from the reactor, and this advantage will increase with distance.

The pure water produced by an on-site hydrogen–air fuel cell will contribute significantly to the total fresh-water requirements of the average household as shown by the following calculation. Burning one mole of hydrogen produces one mole of water and $2 \times 96{,}500$ coulombs. Since the fuel cell is 75 % efficient, the cell voltage will be 0.75×1.23 V. Therefore, burning one mole of hydrogen produces

$$0.75 \times 1.23 \times 2 \times 96{,}500/1000 \times 3600 = 0.049 \text{ kWhr/mole}$$

Hence, the production of 50 kWhr produces 50/0.049 moles of water as a by-product, i.e., 18.2 liters per day.

| 7 | The overall length and width of a typical compact automobile are about 410 and 160 cm, respectively. Assuming that skillful design permits all of this area to be covered by solar cells, a total collection area of 6.56×10^4 cm² could be anticipated.

The total energy collected over a 12 h period of sunshine of average power density, 50 mW cm^{-2}, using photocells of 12 % efficiency would be

$$6.56 \times 10^4 \times 0.050 \times 12 \times 0.12/1000 = 4.73 \text{ kWhr}$$

Taking into account the efficiency of the storage batteries, the total energy available to drive the vehicle in a 24 hr period would be

$$4.73 \times 0.80 = 3.78 \text{ kWhr}$$

Consequently the vehicle could be expected to have a daily range of about 19 miles. While this range falls somewhat short of the average daily distance traveled by commuter automobiles, it is apparent that such vehicles could be of considerable value in many smog-endangered communities. At the present time, however, the cost of solar cells is prohibitively high owing to lack of research and development in the field.

| 9 | The efficiency of a modern coal-burning power plant is about 34 %. Since 1 kWhr is equivalent to 3410 Btu, the total electrical power requirement of the country of 100×10^6 people each consuming 3650 kWhr per year is

$$100 \times 10^6 \times 3650 \times 3410 \text{ Btu/year}$$

Since the efficiency of the power plant is 34 %, the amount of coal required by the country is

$$100 \times 10^6 \times 3650 \times 3410/13{,}000 \times 2000 \times 0.34 = 1.41 \times 10^8 \text{ tons/year}$$

Therefore, the amount of sulfur burned in the process of electrical power generation is 1.41×10^6 tons/year, which produces

$$1.41 \times 10^6 \times 64/32 = 2.82 \times 10^6 \text{ tons } SO_2/\text{year}$$

and the amount of sulfuric acid that could be produced as a by-product of electrical power generation is:

$$1.41 \times 10^6 \times 98/32 = 4.32 \times 10^6 \text{ tons } H_2SO_4/\text{year}$$

This production rate of sulfuric acid corresponds to a yearly per capita production of 86 lb, or about 29 % of the requirements of a highly industrialized country.

For the oxidation of sulfur dioxide (H_2SO_3 in aqueous solution) to be carried out in the fuel cell mode, it must be coupled with a suitable cathodic reaction which takes place at the other electrode. Oxygen reduction is clearly the best choice in this case, as it leads to the formation of no undesirable by-products. Further, the use of oxygen in the air means there are no additional fuel costs, and the high oxygen reversible potential (1.23 V, RHE) results in a high cell voltage and power density. The oxygen electrode, however, does suffer from the drawback that its exchange current density is low and considerable polarization loss usually occurs at this cathode.

The oxidation of H_2SO_3 at the anode proceeds according to

$$H_2SO_3 + H_2O \rightarrow H_2SO_4 + 2H^+ + 2e^- \qquad (11.9.1)$$

while the reduction of oxygen at the cathode proceeds according to

$$\tfrac{1}{2}O_2 + 2H^+ + 2e^- \rightarrow H_2O \qquad (11.9.2)$$

By summing (11.9.1) and (11.9.2), we obtain the overall cell reaction

$$\tfrac{1}{2}O_2 + H_2SO_3 \rightarrow H_2SO_4 \qquad (11.9.3)$$

Since oxygen is in its standard state $[\Delta G_f°(O_2) = 0]$, the standard free-energy change of (11.9.3) is given by

$$-177 - (-129) = -48 \text{ kcal mole}^{-1}$$

where the negative sign indicates that the process proceeds spontaneously with a cell reversible voltage given by

$$4800 \times 4.18/2 \times 96,500 = 1.04 \text{ V}$$

Further, since hydrogen ions do not appear in Eq. (11.9.3), it means that the overall cell voltage is independent of pH; the potentials of both the anode and the cathode vary with pH in the same way. The calculation assumes the partial pressure of oxygen is one and the concentrations of H_2SO_4 and H_2SO_3 are equal.

The cell voltage of 1.04 V is the reversible cell voltage and is not attainable under practical conditions, due to polarization at each electrode and resistance losses inside the cell. The polarization loss at an electrode depends on the current density and electrocatalytic activity of the electrode material for the reaction taking place. Polarization at the sulfurous acid electrode, under practical conditions, is an unknown quantity, but it is probably the same magnitude as that at a working oxygen electrode. This being the case, we could anticipate a working cell voltage of about 500 mV. The generation of electrical power, as well as the sale of the sulfuric acid produced, would seem to make the removal of SO_2 from coal-burning power-plant flue gases economically attractive, for it would produce electricity as a by-product.

Sulfur dioxide in the atmosphere is known to constitute a considerable health hazard and has been shown to promote respiratory ailments in man. Its effect on the environment is hardly less alarming.

Sulfur dioxide stunts and blights natural vegetation and is responsible for extensive property damage by hastening the corrosion of metals, the dissolution of marble, and the weathering of protective and decorative coatings. Of course, not all the sulfur dioxide in the atmosphere comes from coal-burning power plants, but it is difficult to understand why sulfur dioxide is not scrubbed from flue gases and converted into commercially valuable sulfuric acid.

CHAPTER 12

MISCELLANEOUS PROBLEMS

1 The symmetry factor β is a central one in electrode kinetics. Deduce its relation for a general multistep reaction to the transfer coefficient α and to the stoichiometric number ν.

Draw diagrams to illustrate the circumstances under which you would expect to attain "barrierless discharge," i.e., a rate at which there is no longer any effect of overpotential on the reaction rate.

2 Benzoic acid (which can be considered relatively undissociated, and hence, for the purposes of the problem, a nonelectrolyte) is in an aqueous solution at saturation. The solution is now made $1\,M$ in NaCl. On the assumption that the appropriate model is that of ions solvated by water molecules and that water molecules so restricted cannot aid the solution of benzoic acid, calculate the change of solubility of the nonelectrolyte.

The NaCl is replaced by $N(Pr)_4Cl$. What qualitative change would be expected in the solubility of the benzoic acid and why?

3 In ellipsometry, it is assumed that the refractive index of the metal is independent of the potential of the electrode. On this basis, changes in optical properties of the surface are regarded as occurring entirely due to the formation of films thereon. However, it has been suggested that when the electrode potential changes, it is tantamount to a change of electron concentration in the surface of the electrode and therefore to a change in the refractive index.

Review the recent literature on this topic and on this basis estimate

167

numerically the extent to which electromodulation at the surface interferes with the ellipsometric determination of adsorption.

4
To pump blood around, the body needs a power source of about 20 W. Carry out calculations to see whether an internal fuel-cell-powered electrical pump could be a replacement for the human heart, the fuel being provided by the body. What would be the probable anodic and cathodic reactions? What would be the orders of magnitude of the electrode areas involved? (Calculate this from reasonably assumed values of exchange current, etc.) Where could the electrodes be placed in the body? What competing reactions may interfere and could they be electrochemically controlled? What would be the effects of local pH changes arising from the reduction of O_2 in the blood (neutral pH) and from the oxidation of some substances?

5
Consider the following well-known reactions:

$$SO_2 + \tfrac{1}{2}O_2 \rightarrow SO_3 \qquad (12.5.1a)$$

$$SO_3 + H_2O \rightarrow H_2SO_4 \qquad (12.5.1b)$$

$$NO + \tfrac{1}{2}O_2 \rightarrow NO_2 \qquad (12.5.2a)$$

$$2NO_2 + \tfrac{1}{2}O_2 \rightarrow N_2O_5 \qquad (12.5.2b)$$

$$N_2O_5 + H_2O \rightarrow 2HNO_3 \qquad (12.5.2c)$$

$$N_2 + 3H_2 \rightarrow 2NH_3 \qquad (12.5.3)$$

$$N_2 + O_2 \rightarrow 2NO \qquad (12.5.4a)$$

$$2NO + 4H_2 \rightarrow N_2H_4 + 2H_2O \qquad (12.5.4b)$$

Suggest fuel cell possibilities for each of these reactions by writing appropriate anodic and cathodic reactions. Estimate standard free-energy changes for the overall processes occurring in appropriate media at 25°C. For processes that are thermodynamically unfavorable at this temperature, repeat the estimations for higher temperatures within ranges which you think are practical.

Use solubility and diffusion coefficient data to estimate limiting current densities at each electrode and, by analogy to similar processes, estimate the activation overpotential at the limiting current density.

By examination of the literature, estimate a common multiplying factor for the current density in going from planar to porous electrodes. In this way, very roughly estimate the power density of actual fuel cells based on the above reactions. Also calculate the rate of product formation per square meter of electrode and discuss the practicality of any spontaneous electrolytic syntheses which seem to be suggested by your calculations.

6 | One mechanism for the electrocatalytic reduction of O_2 is said to involve the rate-determining step

$$O_2 + H^+ + e^- \rightarrow OH_{ads} + O_{ads} \qquad (12.6.1)$$

Consider from a quantum-mechanical viewpoint the factors which affect the velocity of this reaction at an electrode–solution interface. By what criterion can one know if the process is likely to be adiabatic or nonadiabatic? Draw qualitative potential energy profile diagrams for the reaction pathway orthogonal to the electrode surface. How will the reaction rate depend on the Fermi level in the metal? On the bonding to the substrate? What will be the state of O_2 binding to the electrode?

7 | The analysis of transient data, after the double-layer charging is complete, is used to determine the concentration of radicals on the surface and the rate constants of non-rate-determining steps. The variation of the observed current or potential generally lasts in the millisecond region. However, for the oxidation of a number of organic compounds, the time variation of the current of a potentiostatic transient lasts up to tens of minutes. The interpretation of the early part of the transient has been made in terms of diffusion to the surface. Discuss the time dependence of the oxidation of unsaturated hydrocarbons from this viewpoint assuming that the discharge of a water molecule is the rate-determining step and that the coverage by organic species is high. Show that at constant potential

$$d[\ln(-di/dt)_V]/dt = -D/K\delta \qquad (12.7.1)$$

where D is the diffusion coefficient, δ the diffusion layer thickness, and K the distribution constant of the organic between the electrode and the solution. How would you obtain K and thus examine the applicability of this equation?

8 In the early 1970's, several types of electrochemical power sources seem to hold promise of developing into nonpolluting and non-CO_2 emitting power sources suitable for automobiles. They may be broadly classified as follows:

(a) High energy-density power storers.

(b) Fuel cells burning nonhydrocarbon synthetic fuels, e.g., hydrazine.

(c) Fuel cells burning hydrogen.

Of course, there are also some negative aspects associated with each type of power source. Review these possibilities as quantitatively as possible in the light of the latest information available in the literature and decide which seems to be the optimal system to concentrate developmental work on at the present time.

9 A problem of antipollution technology is the removal of foreign matter from lakes. It has been suggested that much of this material is colloidal and it may be possible to flocculate it electrically (Kuhn, 1971) by evolving oxygen bubbles below it, and floating the coagulated solids to the surface for burning, sweeping away, etc.

Perform appropriate calculations to examine this concept. What concentration of colloidal matter is typical of polluted lakes such as Michigan? What area and type of electrode structure could be envisaged for the achievement of Kuhn's concept? What sort of field and current would be needed? Could significant quantities of a Great Lake be revived in this fashion in ten years? What order of magnitude of expenditure would be involved in capital investment and in operating costs?

10 What is the Esin–Markov effect and how is it related to contact adsorption? Show that the isotherm derived from the BDM theory of the double layer is quantitatively consistent with the Esin–Markov effect for smaller contact-adsorbed ions, e.g., Cl^-, Br^-. Why do inconsistencies appear for larger anions?

11 A solution consists of 0.1 M Cu^{++} and 0.3 M Ni^{++}. It is desired to recover 99.9 % of the copper in its most desirable form, i.e., as bars or strips of metal. Carry out the necessary calculations to act as a basis for an engineering design of a continuous copper extractor. Consider the transport properties and the deposition potentials of the ions in this solution and estimate the exchange current densities for the deposition of each. Decide on the potential, type of electrode, flow rate, and type of flow (laminar or turbulent) of the solution that would lead to optimal recovery of pure copper. The design of the electrode should be compatible with minimal cell volume.

Derive an equation for the fraction of copper removed in one pass of the solution through the extractor and hence for the extraction time as a function of flow rate. Calculations of the optimal electrode potential and flow rate should be performed and the cell voltage and efficiency estimated. What other considerations are important?

[12] Review the physical and chemical properties of the sodium tungsten bronzes and the recent literature pertaining to their use as electrodes. Discuss these aspects of bronze electrodes:

(a) Evidence which may lead to the sodium tungsten bronzes being considered as semiconductor electrodes.

(b) The promotion by Pt of the catalytic activity of the bronzes for oxygen reduction. Is it the special properties of traces of Pt on the surface of the bronzes or the properties of bronzes doped by Pt which is important?

(c) The value of studies on the bronzes to the future development of oxygen electrodes.

(d) The difficulties associated with the preparation of porous electrodes from the bronzes and how such difficulties might be overcome.

(e) The economic aspects of sodium tungsten bronze electrodes containing a few hundred ppm of platinum for oxygen reduction and evolution.

APPENDIX 1

VALUES OF SOME IMPORTANT PHYSICAL CONSTANTS

Avogadro's number, $N_A = 6.023 \times 10^{23}$ mole^{-1}

Boltzmann constant, $k = 1.381 \times 10^{-16}$ ergs deg^{-1}

Electron charge, $e_0 = 4.803 \times 10^{-10}$ esu

Electron rest mass, $m_e = 0.101 \times 10^{-28}$ g

Faraday constant, $F = 96,490$ coulombs mole^{-1}

Gas constant, $R = 8.314$ J deg^{-1} mole^{-1}
$= 1.987$ cal deg^{-1} mole^{-1}

Mechanical equivalent of heat, $J = 4.184$ J cal^{-1}

Planck constant, $h = 6.626 \times 10^{-27}$ erg sec

Triple point of water, $T_0 = 273.16°$K

Velocity of light, $c = 2.998 \times 10^{10}$ cm sec^{-1}

Volume of perfect gas (NTP) $V_0 = 22.414$ liters mole^{-1}

APPENDIX 2

SOME USEFUL CONVERSION FACTORS

1 ft^3	= 28.32 liters
1 atm	= 1.013 $\times$ 10^6 dynes cm^{-2}
1 Btu	= 1054 J
	= 0.2929 Whr
1 kWhr	= 36 $\times$ 10^5 J
1 eV	= 23.06 kcal mole^{-1}
1 coulomb	= 3 $\times$ 10^9 esu (statcoulombs)
1 A	= 3 $\times$ 10^9 statamperes
1 V	= 0.333 $\times$ 10^{-2} statvolt
1 ohm	= 0.111 $\times$ 10^{-11} statohm
1 F	= 9 $\times$ 10^{11} cm
1 esu cm^{-1}	= 1 statvolt

BIBLIOGRAPHY

Beck, W., J. O'M. Bockris, J. McBreen, and L. Nanis, *Proc. Roy. Soc.*, **A290**:220 (1966).

Bockris, J. O'M., *The Electrochemistry of Cleaner Environments*, Plenum Press, New York (1971).

Bockris, J. O'M., and D. Drazic, *Electrochemical Science*, Taylor and Francis, London (1972).

Bockris, J. O'M., and Z. Nagy, *Ecological Electrochemistry* (1972) [projected date of publication, 1973].

Bockris, J. O'M., and A. K. N. Reddy, *Modern Electrochemistry*, Plenum Press, New York (1970).

Bockris, J. O'M., and P. P. S. Saluja, *J. Phys. Chem.*, **76** (15):2140 (1972).

Bockris, J. O'M., and S. Srinivasan, *Fuel Cells—Their Electrochemistry*, McGraw-Hill, New York (1969).

Bockris, J. O'M., N. Bonciocat, and F. Gutmann, *A Primer in Electrochemical Science*, Wykeham Press, London (1972) [publ. 1973].

Booth, F., *J. Chem. Phys.*, **19**:391 (1951).

Brummer, S. B., J. I. Ford, and M. J. Turner, *J. Phys. Chem.*, **69**:3424 (1965).

Conway, B. E., *Electrochemical Data*, Elsevier, New York (1952).

Conway, B. E., *Proc. Roy. Soc.*, **A247**:400 (1958).

Conway, B. E., in *Modern Aspects of Electrochemistry*, Vol. 3, Ed. by J. O'M. Bockris and B. E. Conway, Plenum Press, New York (1964).

Conway, B. E., *Electrode Processes*, Ronald Press Co., New York (1965).

Conway, B. E., J. O'M. Bockris, and H. Linton, *J. Chem. Phys.* **24**:834 (1956).

Criss, C. M., and J. W. Cobble, *J. Amer. Chem. Soc.*, **83**:3223 (1961).

Delahay, P., *J. Amer. Chem. Soc.*, **75**:1190 (1953).

Delahay, P., *New Instrumental Methods in Electrochemistry*, Interscience, New York (1954).

Delahay, P., *The Double Layer and Electrochemical Kinetics*, Interscience, New York (1965).

Eigen, M., and L. deMaeyer, *Proc. Roy Soc. (London)*, **A274**:505 (1958).

Emi, T., and J. O'M. Bockris, *J. Phys. Chem.*, **74**:159 (1970).

Eyring, H., D. Henderson, and W. Jost, *Physical Chemistry—An Advanced Treatise*, Vols. 9A and 9B, Academic Press, New York (1970).

Falkenhagen, H., and W. Ebeling, in *Ionic Interactions*, Ed. by S. Petrucci, Academic Press, New York (1971).

Frank, H. S., in *Chemical Physics of Ionic Solutions*, Ed. by B. E. Conway and R. G. Barradas, John Wiley and Sons, New York (1966).

Frank, H. S., and P. T. Thompson, in *The Structure of Electrolyte Solutions*, Ed. by W. Hamer, John Wiley and Sons, New York (1960).

Fricke, H., *Phys. Rev.*, **24**:575 (1924).

Friedman, H. L., in *Modern Aspects of Electrochemistry*, Vol. 6, Ed. by J. O'M. Bockris and B. E. Conway, Plenum Press, New York (1971).

Frumkin, A. N., *Z. Phys.*, **35**:792 (1926).

Frumkin, A. N., and B. B. Damaskin, in *Modern Aspects of Electrochemistry*, Vol. 3, Ed. by J. O'M. Bockris and B. E. Conway, Plenum Press, New York (1964).

Fuoss, R. M., and F. Accasina, *Electrolytic Conductance*, Interscience, New York (1959).

Gileadi, E., G. Stoner, and J. O'M. Bockris, *J. Electrochem. Soc.*, **113**:585 (1966).

Goffredi, M., and T. Shedlovsky, *J. Phys. Chem.*, **71**:2176 (1967).

Gregory, D. P., D. Y. L. Ng, and G. M. Long, in *The Electrochemistry of Cleaner Environments*, Ed. by J. O'M. Bockris, Plenum Press, New York (1971).

Halliwell, H. F., and S. C. Nyburg, *Trans. Faraday Soc.*, **58**:1126 (1963).

Harned, H. S., and B. B. Owen, *The Physical Chemistry of Electrolyte Solutions*, Reinhold, New York (1958).

Hemmes, P., and S. Petrucci, *J. Phys. Chem.*, **72**:3986 (1968).

3638 ⸺

Henderson, P., Z. Physik. Chem., **59**:118 (1907).

Henderson, P., Z. Physik. Chem., **63**:325 (1908).

Horiuti, J., and M. Polanyi, Acta Physicochem. URSS, **2**:505 (1935).

Kovac, Z., Dissertation (University of Pennsylvania) (1964).

Kuhn, A. T., H. Wroblowa, and J. O'M. Bockris, Trans. Faraday Soc., **63**:1458 (1967).

Latimer, W. M., Oxidation Potentials, Prentice-Hall, Englewood Cliffs, New Jersey (1952).

Levich, V. G., in Physical Chemistry—An Advanced Treatise, Vol. 9B, Ed. by H. Eyring, D. Henderson, and W. Jost, Academic Press, New York (1970).

Marcus, R. A., J. Phys. Chem., **67**:853, 2889 (1963).

Nanis, L., and J. O'M. Bockris, J. Phys. Chem., **67**:2865 (1963).

Nagarajan, M. K., and J. O'M. Bockris, J. Phys. Chem., **70**:1854 (1966).

Nightingale, Jr., E. R., J. Phys. Chem., **63**:1381 (1959).

O'Ferrall, R. A. More, G. W. Koeppl, and A. J. Kresge, J. Amer. Chem. Soc., **93**:1 (1971).

Parsons, R., Trans. Faraday Soc., **147**:1332 (1951).

Parsons, R., Handbook of Electrochemical Constants, Butterworths, London (1959).

Passynski, A., Acta Physicochem. URSS, **8**:385 (1938).

Planck, M., Ann. Physik., **39**:161 (1890a).

Planck, M., Ann. Physik., **40**:561 (1890b).

Pourbaix, M., Atlas D'Equilibres Electrochimiques, Gauthier-Villars, Paris (1963).

Pourbaix, M., Lectures on Electrochemical Corrosion, Plenum Press, New York (1973).

Rampolla, R. W., R. C. Miller, and C. P. Smyth, J. Chem. Phys., **30**:566 (1959).

Sand, H. J. S., Phil. Mag., **1**:45 (1900).

Solomons, C., J. H. R. Clarke, and J. O'M. Bockris, J. Phys. Chem., **49**:455 (1968).

Vetter, K., Electrochemical Kinetics, Academic Press, New York (1967).

Zachariasen, W., J. Amer. Chem. Soc., **54**:3841 (1932).

Zana, R., and E. Yeager, J. Phys. Chem., **71**:521 (1967).